T4-AKD-417

Effective TECHNICAL WRITING & SPEAKING

Barry T. Turner
MSc, CEng, FIMechE, MICE, MIProdE, AFRAeS, MBIM
Professor of Industrial Management, University of Newcastle-upon-Tyne

COLES PUBLISHING COMPANY LIMITED
TORONTO – CANADA
PRINTED IN CANADA

Originally published in United Kingdom - which accounts for British references.

Contents

Preface

The second edition of this book, like the first, has been written for engineers and scientists—mainly for those who have been working in industry for a few years. However it is hoped that it will also be useful to those who are about to embark on any technical career.

Experience has shown that most engineers and scientists underestimate the magnitude of the problems concerned with presenting technical information. The problems are not confined to conveying factual statements or even to matters of grammar or the mastery of spelling or good diction. Rather they are concerned with clear logical thinking—how to organise material for easy comprehension and use. To do this requires skill in three main areas. First, the engineer or scientist must know his subject thoroughly. Secondly, he must recognise the readers interests and needs and, thirdly, he must design and structure his information to serve the reader in the best possible way.

Studies have shown that engineers fail to read widely or even read about their own subjects very much. There is a natural connection between reading and thinking and writing and speaking. The failure of engineers to read probably contributes to their poor competence in the field of technical exposition. The pure scientist however considers it absolutely vital to know what others are doing in his field and it is for this reason that most advance in the selective dissemination of information has occurred in science rather than engineering. Indeed it could be said that science is a culminating activity leading to publication whereas technology does not end in literature but hardware of one sort or another. There is, of course, a considerable body of technological literature but it does not feed on itself in order to grow as science does. One branch of technological literature besides the learned society papers of professional bodies are the vast number of patents published. These record inventions and innovations form a valuable part of technical literature. However the main body of technical writing required in industry today covers specifications, contracts, tendering documents, reports etc. as well as the more popularly styled—and therefore more difficult—commercial and training literature dealing with technical subjects.

Educationalists have confined themselves, in the main, to teaching good grammar and style. Their method of teaching has often deterred all but the most able school pupils. The result is that many numerate pupils develop an aversion to verbal expression, both written and oral, leaving such matters to arts graduates.

Unfortunately there have been several famous engineers and scientists who have contributed much in the way of original research but who appeared to have totally lacked the art of exposition. There is the apocryphal story about William Thompson, later Lord Kelvin, who, it was said, could not lecture well. He was appointed Professor of Natural Philosophy in Glasgow at the tender age of 22. Under him was a lecturer named Day. When Thompson was summoned to London to be created Sir William, the undergraduates wrote on the blackboard, 'let us work while it is yet Day, for the Knight cometh, when no man can work'.

By all accounts, James Clark Maxwell and Osborne Reynolds were both inarticulate and poor at exposition. Doubtless in those days, in the sheltered academic world of the time, their lack of ability to communicate effectively did not matter too much and possibly they had others who could do it better for them. Today, however, we live in a very different climate where technology has imposed change. Such change is accelerating and in the highly competitive world of modern business and industry the need for effective transfer of information to and from research, development, design, production and customers can mean the difference between success and failure in the market place.

While technical facts and important findings may 'speak for themselves' they will only do so when given the opportunity and even then there is no guarantee that they will be heard and understood. Effective technical writing and speaking to communicate the results and implications of the work of engineers and scientists requires some training. In short the art of exposition needs teaching and practising, for all classes of technical personnel, if ideas are to be turned into products and projects speedily with the minimum of wasted effort and materials.

In today's industrial climate there is yet another factor that makes good communications essential. There is an increasing demand for shop-floor participation in management decision-making. It is in the interest of engineers that such decisions shall at least be well-informed. But how can they be that if engineers fail to put across their problems and ideas?

No longer can society tolerate inarticulateness in modern technologists and indeed all professions require some practice of rhetoric which may be regarded as the present day version of the Greek art of communicating knowledge persuasively.

Provided engineers and scientists are prepared to invest time and energy they can improve their writing and speaking abilities. This book is aimed at revealing on what they must concentrate and how they should go about improving their skill as expositors.

A final chapter has been added to this new edition to give a round-up on some of the repercussions of modern communication media on our man-machine society. An attempt has been made to show possible trends in the communication of technical information.

1 The importance of the written and spoken word

The communication of ideas is of vital importance to mankind and essential to human activity. When writing was invented the foundation of civilisation was laid and the miracles that verbal thinking have wrought are indeed great. The raw ingredients of thought are words. Words underline our whole life. They are the signs of humanity, the tools of business, the expression of our affections and the records of our progress. It is the ability to communicate with words that has made man the dominant species on the planet Earth. His tongue and pen have been the interpreters of his mind. No flash of technical insight is worth anything unless it is communicated to others.

Indeed the history of communication is the history of the growth of civilisation. The Bible reminds us that we are all members one of another and the more effectively we communicate the better the quality of life.

Technology has provided various means of long-distance communication from rockets and aircraft to telephones and television. All of these bring people closer together and it is important to recognise that in considering technical writing and speaking the means or media through which they are expressed must be taken into account as well as the matter communicated. Contents of the media cannot be separated from the technologies of the media themselves [1]*. How the media say something, and to whom they say it, affects what they have to say. As the Canadian Professor Marshall McLuhan has pointed out in his *Understanding Media* [1, 2] a statement made on television is very different in both its character and its effects from a statement of the same facts through a newspaper. The images, the spontaneity, the wide dissemination of a television broadcast actually modify the material given.

* Numeral references refer to the references given in Appendix 1. Letter references refer to those given in 'Further Reading' at the end of each chapter.

This means that the expositor needs to know something of a medium's operation and effects if he is to appreciate its likely impact on an audience.

In business there is no inefficiency so serious as that which arises from poverty of language. If a technologist or technician does not express himself meaningfully and clearly he quickly becomes a nuisance wasting everyone's time. The more complex a business becomes the greater the need for clear, concise, unambiguous communication. Every managerial or supervisory job requires the best possible spoken and written English to ensure the fullest understanding between individuals if efficient and effective operations are to be performed. Sir Ernest Gowers [3] gave profound advice when he wrote, 'Be simple, be short, be human, be correct in your written and spoken words.' In conveying technical ideas it is sometimes difficult to choose the right word but most technical writing would improve if familiar and precise words were chosen. In all communication one of the greatest problems to overcome is the illusion that it has been achieved successfully. This is particularly true in the realm of words—both spoken and written. Remember what the late President Kennedy said about Sir Winston Churchill:

> 'In the dark days and darker nights when Britain stood alone, and most men save Englishmen despaired of England's life he mobilised the English language and sent it into battle. The incandescent quality of his words illuminated the courage of his countrymen.'

This was mainly said about Sir Winston's spoken words which motivated the nation and guided the House of Commons through the dark days of war, but it is also true of his writings. We need to remember that Mussolini held Italy in his hand by the power of his words. Austria was conquered by Hitler with words. Indeed it could be said that personal advancement in life lies in the ability to say the right kind of words in the right way at the right time.

1.1 Communication and what it is

Communication is not merely a matter of language—painting and music are both methods of communication. 'Communication' is a highly ambiguous term that can be traced back to the Latin word *communicare* which means 'to share', 'to make common', 'to impart'. It therefore implies that communication is the art of transferring or sharing ideas, information, instructions or feelings. It is a social process being both interactive and purposeful. The basic elements of good communication are clear thinking, clear speaking, and clear writing. Like the word 'sovereignty' or 'freedom', the word communication has a large territory of coverage, for people communicate information, knowledge, error, opinion, ideas, experiences, moods, thoughts, emotions, wishes, orders and so on. But heat and motion can be communicated as can disease. A system of communications covers transmittal from one thing to another or one person to another. Sometimes the word refers to what is transmitted (the message) and sometimes to the means of transmittal (the medium) and sometimes to the whole process. In most applications what is transmitted is shared. If a transmitter conveys information to another person he does not necessarily lose what he has—and here there is a sense of participation. Of course, information can be transmitted of which the transmitter is unaware, as in a behaviour pattern enabling others to know the sender better than he does himself.

In this book we are particularly concerned with the communication of technical information of one sort or another. This kind of communication is really connected with a process of mutual understanding, by which knowledge is shared. In particular, 'selling' involves technical communication, either of an idea or of details of a product. Here success is measured not necessarily by understanding but by the overt response of the recipient. In management, 'informing' involves communication, e.g. a supervisor communicating with his or her subordinates in a hierarchical system that imposes authority. Here, too, the right response is the test of good communication.

Nowadays management communications are not confined to business matters: in many countries there exist stringent laws and regulations regarding the necessity for effective communication of safety requirements of a job.

1.2 Business and communication

Business is becoming more and more complex: organisations are tending to group together and to increase in size; establishments in the same group of companies may frequently be separated geographically by considerable distances. Often, therefore, it is impossible to hold effective meetings or to contact all interested parties. Consequently the written word has to be used to inform. Furthermore, as business becomes more complex and products more diversified and sophisticated, more knowledge about both is generated. This knowledge has to be accurately recorded for reference and re-use to avoid duplication of effort. Moreover, policies, procedures, agreements and contracts also increase as industry grows and many of these contain legal obligations that have to be met, hence the importance of accurate, clear communication.

To survive, firms today are seeking the best media for storing and expediting information. Much engineering information is coded and stored in a highly stylised form, e.g. drawings, schematics and other diagrams. Other technical information is in the form of reports, data sheets, papers and sundry documents. The former is preserved in visual form; the latter is mainly in written language. Much of business is concerned with processing and movement of information.

In business it is important to recognise that the content of a message changes as it passes from one level of the organisation to another as for example from 'top management' to 'operator level'. In a science-based industry ideas may start in a research department (see Fig. 1) and pass through several phases as they are changed into instructions. At the interface of say research and design there may be considerable communications difficulties, likewise at the design/production interface where the instructions are turned into hardware. As progress is made so the language is changed and a continuing process of translation and interpretation occurs throughout the whole industrial process.

The social factors affecting communication are also important in business. These include such problems as the size of group, the influence of the characteristics of the individuals comprising the groups, and the effect of instruction.

In addition to all these we are increasingly discovering the importance of psychological factors in communication, particularly unconscious ones. For example, if you don't like someone you are less likely to get your ideas across to him/her.

1.3 Management and communication

Technical managers are today looking for men and women who can

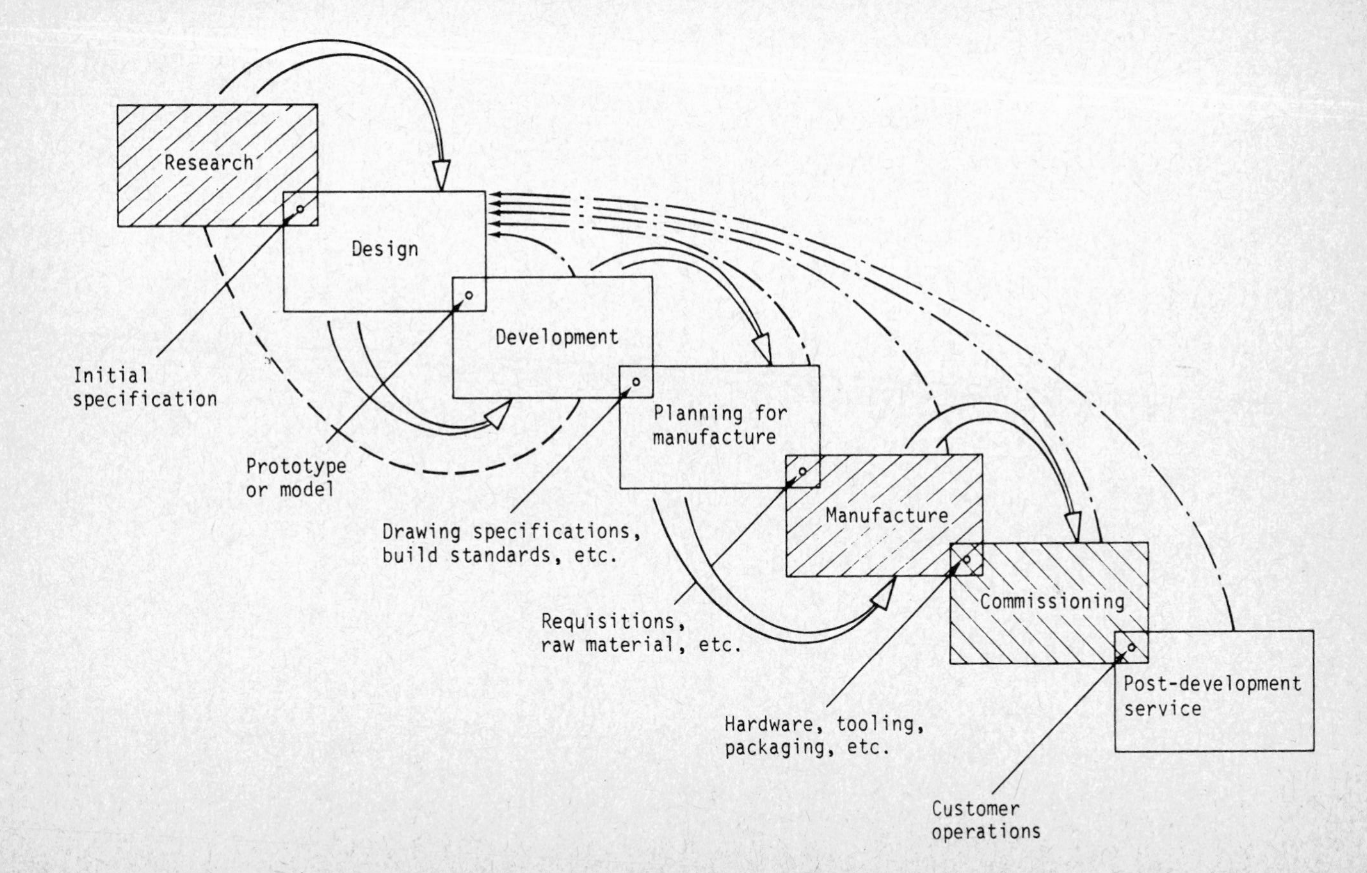

Figure 1 The translation process—from scientific discovery to practical hardware

report quickly on paper the day to day facts of their work. They require more than undigested data. They are seeking an interpretation of the data. Unfortunately individuals attach a wide variety of meanings to words. Often interpretations vary widely, even among groups with similar, specialisations, e.g. 'real time' in the computer world, 'design process' in engineering.

Indeed, management of all kinds has a great responsibility to ensure that adequate communication occurs in their organisation. Chester Barnard pointed out that 'the first executive function is to develop and maintain a system of communication'.

Management must decide upon the form and content in any items of communication but to do this properly it is essential first to decide the exact purpose of this communication and obtain a thorough appreciation of the knowledge, understanding, interests and attitudes of those who will receive the information. This means that communication must be considered as an aspect of human relations. Everyone finds it easier to communicate with people they like and feel at ease with. A writer must keep his reader ever before him as he structures his message. His layout, presentation and style will vary with the reader and types of readers.

Management must think of communication as perception and understanding and take into account the cultural and social climate. Is the message couched within the range of the recipient's perception? Very often the perception limitations are cultural and emotional—not lack of reasoning.

One of the most extensive areas in industry where better management communication is required concerns the transference of information concerning an employee's job. In the U.K.,the Code of Industrial Relations Practice, which has been prepared to implement the Industrial Relations Act clearly puts the onus on management for ensuring effective communication and consultation. It shows too that good communication implies involvement. Individual employees will in future have to be given several specific items of 'essential information' including what is required of them in their jobs and to whom they are responsible; opportunities for promotion and any training necessary to achieve a new level of responsibility. Also information must be given about organisational changes that affect employees. In the technical realm it is not always easy to write job descriptions or terms of reference. When these are done well they provide a useful appraisal means and can help with manpower planning.

Managers must constantly seek to improve their communication performance. They should always question their real objective in each message and examine how it can most effectively be expressed, bearing in mind the language and style best suited to the readers or listeners. They should also consider the most appropriate media of communication. The choice of media will be governed by cost and desired impact to be achieved [2].

1.4 The communication process

Man perceives the world with his five senses but mostly employs only two in communicating. Hearing and seeing supply the information which undergoes a process of storage, collation and selection in the brain and nervous system before it is used. Only by the use of information can man acclimatise himself to his surroundings—living effectively is living with adequate communication.

Long before the days of mass communication and commercials the Psalmist wrote 'Much vain disquiet maketh mankind'. This seems to be even truer today than when it was first penned. More and more 'noise' is being generated within our communication systems.

Today there is a communication explosion. Message traffic has increased all over the world by means of new technical methods. As the transformation of human wishes into forces greater 'than those of muscles is achieved by amplification, so amplification is necessary for communication to extend its range. But in spite of this increased range of communication by telephone, telegraph, radio, television and space satellites, etc., all too often the message itself proves to be inadequate. It is a paradox of modern life that as communications become easier, communication becomes more difficult.

When Marconi, in 1901, carried out his radio experiment across the Atlantic against the better judgement of the scientists of the day, a new dimension was added to social life by the ability of man to cross cultural barriers. Yet communication leading to real understanding is still a major problem of mankind. Looked at from a personal point of view, people are being inundated with information and are finding it difficult to stay afloat in the new ocean of knowledge. This is particularly true of technical knowledge.

All communication systems terminate ultimately in human beings and for auditory signals the receiver is the ear, for visual signals it is the eye. However carefully the emission of messages is purified there is no certainty that such messages will be understood in the way intended. The reason for this is that behind the eye and the ear a complex control system operates in the brain before the recipient can incorporate the message. Moreover, the degree of receptivity depends upon a number of subjective factors which can affect the way in which any message is interpreted.

It is at this stage of the communication process that clear speaking, good voice production, and precise writing are of vital importance. But they will only be so if the recipients are prepared to look and see and listen and hear. Too often the spoken and written word is outside the frame of reference of the audience and above their intelligence. The BBC audience research department reports have shown conclusively that talks at a level of difficulty appropriate to the top third of the population can rarely convey much to people of even average intelligence, and little or nothing to the backward quarter of the population. Furthermore, comprehension of broadcasts is profoundly influenced by the extent to which people are interested in the subject under discussion or have their interest aroused. The greater the listener's interest the greater their understanding is likely to be, and vice versa [4].

Note that successful communication is also achieved by changes or modulation. Without modulation, radio, television, and even human speech would be impossible. For human speech requires the help of the palate, teeth, tongue and lips to modulate the air column from our lungs (see Appendix 7). In the case of pictures ink is placed on paper in such a way that it modulates the incoming visual impressions.

All forms of communication—machine, animal or human—have one common feature, namely, 'signs' which are used according to certain rules.

In the case of language, there are rules grouped together under the titles of semantics and syntax, the first concerned with the meaning of words and the second with the way they are put together into sentences. But numerous other signs, besides words themselves, can be used to convey information: gestures, punched holes in paper tape, electrical impulses and so on. This representation of information in the form of signals is called encoding and is used in some way in all communication. A message is first encoded into signals, then transmitted and finally received. But the accuracy of the final reception depends upon proper decoding.

Codes of many types may be used for representing messages, but most may, if required, be reduced to a simple binary form. There can be no signal, however, without some expenditure of energy.

Only after the last war, in 1948, did Shannon, Wiener and others [B, E] provide us with an adequate mathematical theory of communication. The 'Shannon' model consists of a source of information (input), which may be a machine or a person, and an encoder. These together constitute the transmitter (see Fig. 2). The transmitter is connected to the receiver by means of a channel along which signals are sent. These are decoded on reception for output to another person or a machine. In any real system the channel will be 'noisy'. Noise can be considered to be any random effect which interferes with or masks signals sent. The receiver thus obtains faulty information and the degree of error can be measured. In other words Shannon set out to discover how fast it is possible to transmit a given type of message over

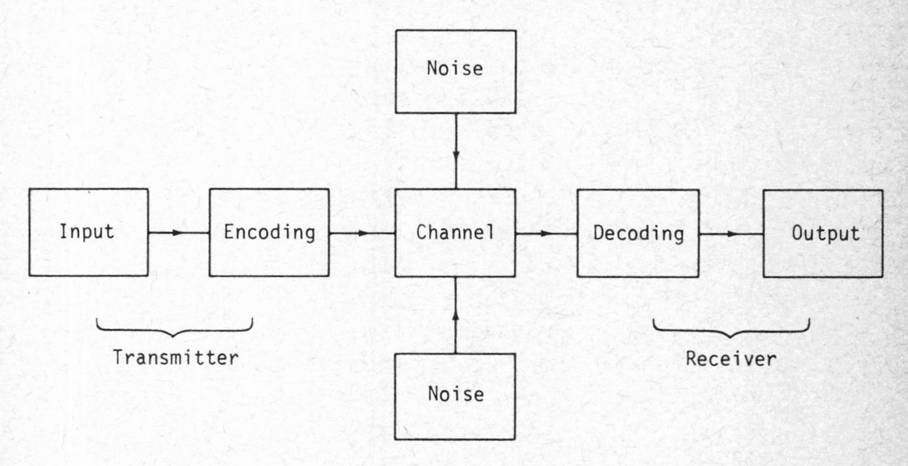

Figure 2 The Shannon model of the communication process: the requirements are modulation, amplification an energy for a message to be sent

a channel of specified characteristics, without error, which enabled man to maximise the speed of error-free transmission of a given type of message over a set channel.

The process of communication requires expenditure of energy, encoding (modulation), amplification, detection and decoding. But with all the increased knowledge of the process and methods of communication real understanding of what is communicated has not increased to the same extent. The 'Shannon' model makes one basic assumption that is valid for inanimate systems but not for the human situation. The assumption is that the encoder of the transmitter and the decoder of the receiver perform exactly inverse operations. But with humans it is unlikely that this will ever be so. Not only is a receiver likely to misinterpret the message but he may well add to it by making selective use of his many sensory inputs. A more realistic model of the human communication system can be represented as in Fig. 3. Here codification is represented by scribing, writing, or speaking, etc., by the sender, be it a man or machine, and energy has to be expended to transmit the signals. Detection and decoding are done by the receiver through his senses and mind. The comparatively simple representation of the communication process belies the main problems. For example, the sender often fails to define his objective clearly—what does he want the reader or listener to know or feel or do? He may fail to organise his material satisfactorily and fail to choose the appropriate words. Being a good sender requires a knowledge, therefore, of what one wants the recipient to know, feel or do. The receiver may, due to past experiences, detect the message but distort it in the decoding process. Often he will tend to misinterpret messages which he does not like e.g. because they threaten his established position. There may also be considerable preoccupation barriers which can block or prevent accurate response to the presented material such as emotions, attitudes and language. In a communication that is broadcast to a large audience, there may well be other members of his group who can affect the individual's interpretation.

While the process in oral and written communication is similar there is an added visual dimension in oral communication: mannerisms, body posture and movement and facial expressions become part of the message. In written communication the receiver may not bother to read further than the first line or put it aside once he meets a difficulty or becomes bored. It could be said that the manner of presentation in an oral communication system is analogous to the style and handwriting or typing and layout in a written communication.

1.4.1 Man-to-man communication

Good communication like good design, begins in the mind. Certain

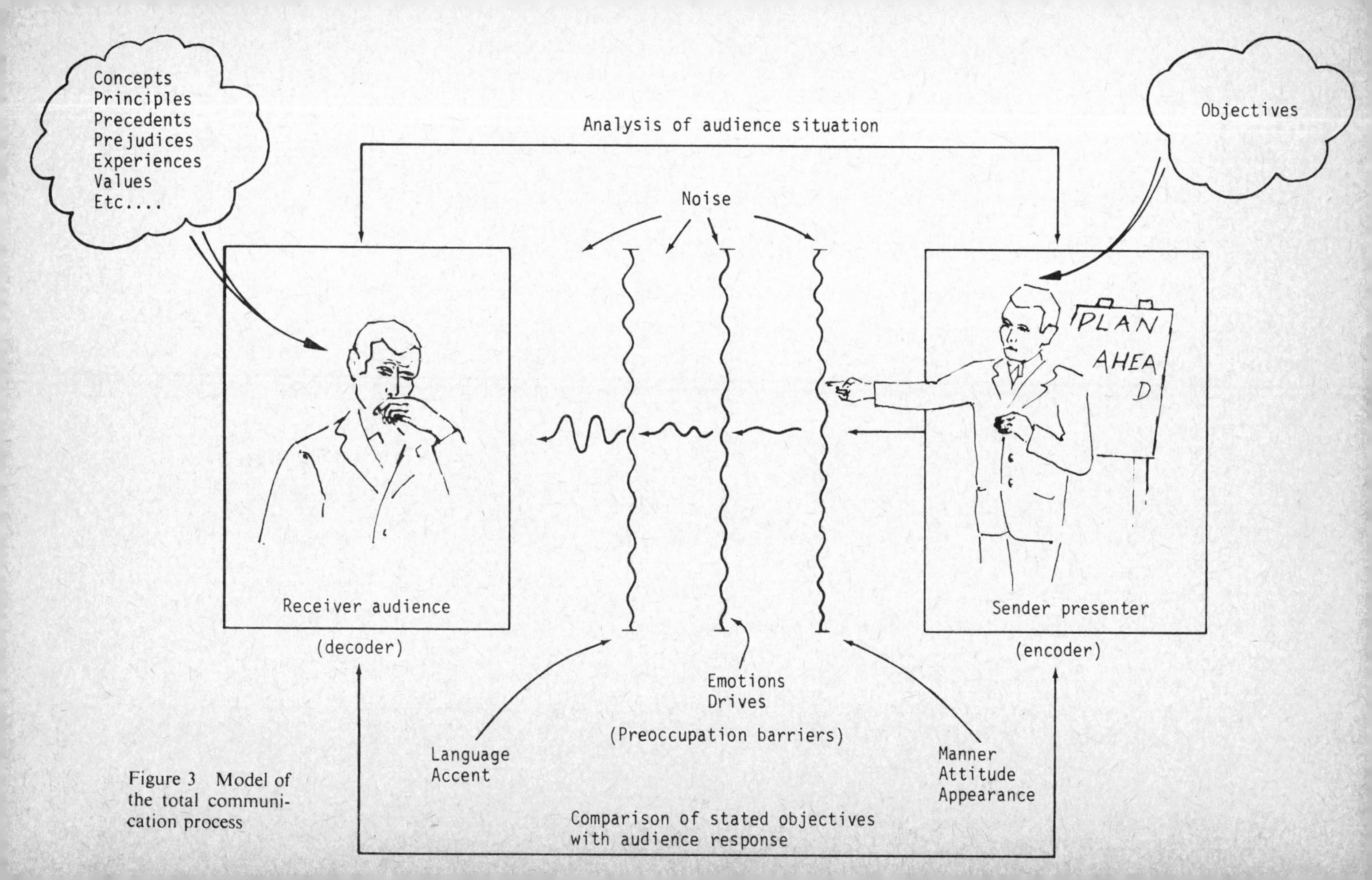

Figure 3 Model of the total communication process

attitudes and thought processes will precede the words and a good communicator will always consider the recipient of his message. On a face-to-face basis the use of mannerisms, intonation of voice and other factors can help to convey meaning. But imparting information requires more than just a series of words and gesticulations. It demands a proper approach to win interest and so a hearing. And even if these factors are carefully considered, man to man communication can still prove difficult because of the differing interpretation of words. In general, a problem arises when both parties fail to recognise a misunderstanding and so both feel completely secure in the midst of their communication breakdown. Getting things across is not so much a matter of how they are expressed as of securing a receptive atmosphere, finding a common viewpoint, or changing an existing one. It is interesting to note that the emotional impact of a face-to-face communication matters far more than what is said and how it is said. For it is only after a reassuring impression has been registered that people are prepared to listen [H, I].

Man's biggest problem lies in languages. Human language, and in particular the concoction of the alphabet, must be classed as man's greatest inventions. Similarly, the Tower of Babel must be one of man's worst disasters. Language is the vehicle for storing and conveying meaning and it does so by means of a large body of unexpressed convention and background. It is man's most important social faculty and sets him above the crude animal sign-systems which do not have the same universal properties [A, G].

Different languages have different properties, e.g. German is a very rigid and logical language and consequently is very ponderous for translating new technical terms.

Chinese written words/symbols originate from pictures. It is claimed that a learned Chinese can memorise 50,000 of them, but ordinary people get along with about 3,000. For sensitive communication in Chinese a reader needs to be able to identify 30,000 picture words. It is all very ponderous and subtle differences occur by building up characters. A 'man' is spelt by a picture of two legs. 'Thunder' is spelt by two legs chasing another two legs. 'Wife' is one woman with a broom, 'gossip' is represented by three women together! The whole process of written communication is tedious and time-consuming and does not lend itself to easy manipulation. Perhaps language colours the way a people acts or does the way it acts become reflected in the language? In

Spain time walks, in England time flies. Certainly language plays an important role in man's ability to think. English is one of the best all-purpose languages since it gives a great deal of freedom compared with say German and all scientific personnel should be grateful for this fact.

It is worth reflecting that all science, literature, and formal logic would be impossible without words. Words are the primary tool in man's domination of this planet. They can be easily produced, reproduced, stored and transported. Man uses words to describe people, rings, actions, or relations between them. In the case of actions, for example in defining length or time one has to give a description for measuring quantities. The measuring procedure is an activity or operation and such definitions connect the verbal and physical worlds. But there are some words which are essentially used for dealing with other words, such as if', 'or', and 'not'. They are tools for carrying out logic processes just as addition, subtraction, multiplication and division are used to process and manipulate quantities. They are particularly relevant to the formulation and communication of ideas.

How do we attach meaning to words? With the word 'red' it seems quite straightforward. When we were children, someone perhaps pointed to a red object and said, 'Red'. We all have roughly the same idea of what red is. But it is difficult to take a child and point to 'justice' or 'immediately'. It seems that the way in which we build up meanings for words is a very subtle and complex process. Small wonder that people have difficulty in agreeing over what is meant by a 'fair wage'. Take the word 'run'. In engineering we talk about 'a run of work', or 'a weld run'. In everyday life people speak of 'The stream runs swiftly' or a fence is said 'to run around a building'. None of these uses connotes leg action as in 'running a race'. Originally they were all metaphors so today we take the word to mean that which all its applications have in common, namely describing a course. Metaphor is a characteristic of language growth [J].

An engineer can use many technical words, of which only a few may be intelligible to the layman. It is essential, therefore, to determine the appropriate level of vocabulary to be used when communicating. So often, technical jargon is used where Anglo-Saxon words would be adequate. In this connection the English word 'pyramid' shown in Fig. 4 at (A) indicates that 900 to 1000 words would cover about 85 per cent of a writer's requirements on all ordinary subjects. Obviously

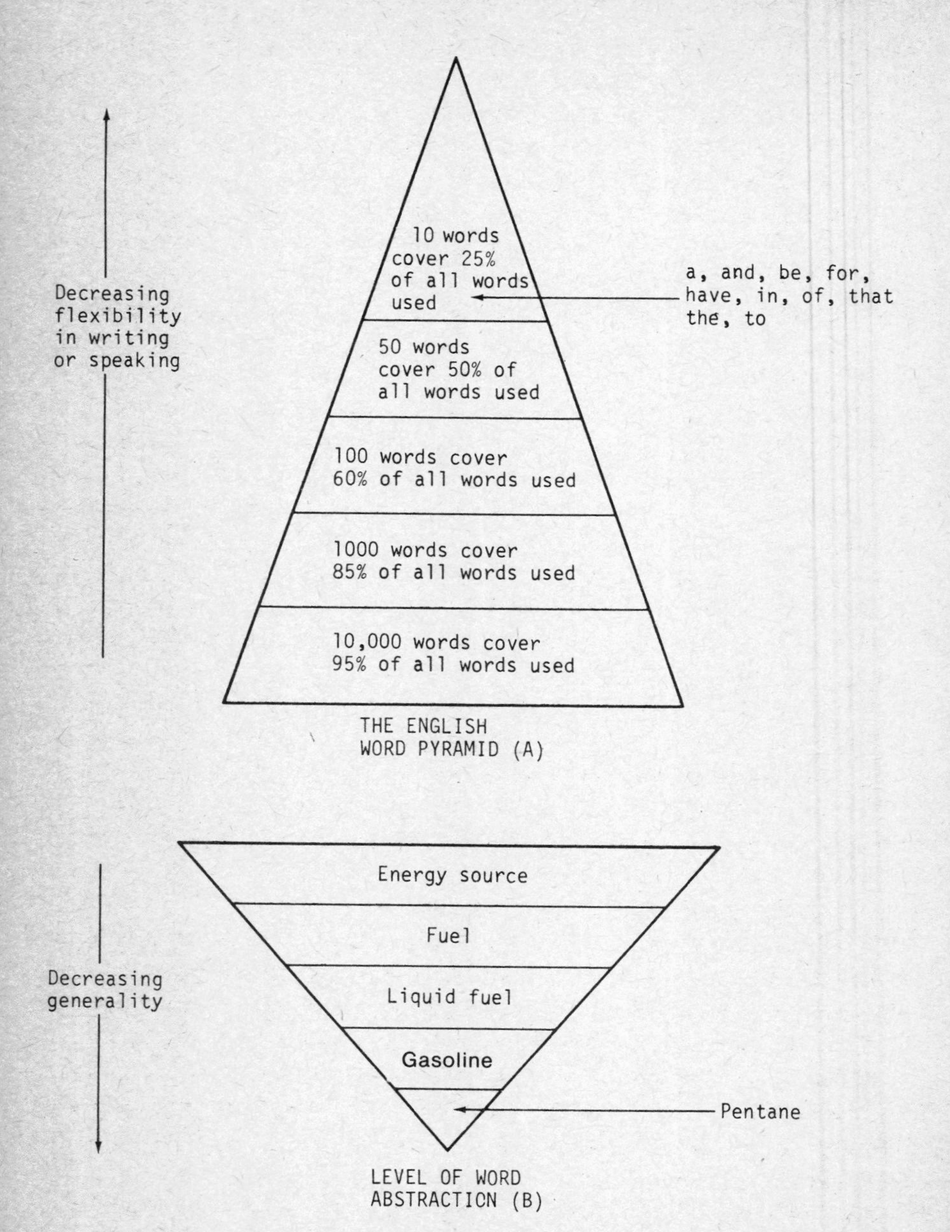

Figure 4 The English word pyramid and the level of word abstraction

technical terms are convenient for technical writing but where necessary they may need to be defined.

In chemical engineering or in the physics laboratory engineers and scientists are mainly concerned with tangible objects, operations and processes that can be demonstrated. Social scientists are not so fortunate in this respect and find it difficult to keep their language at a concrete level. This is why some of their findings are not understood by engineers and scientists. In general it may be said that there is a hierarchy of abstraction as shown in Fig. 4 at (B). The terms decrease in generality as we move down from energy source to pentane. It is a good rule when writing or speaking to keep to the lowest level of abstraction possible. In other words don't move up the abstraction ladder further than necessary to convey the correct sense.

For all the importance of words, we must not confuse them with the things they represent. Words are only symbols. The American city that banned all publications containing the words Lenin and Leningrad did little to restrain Communism. But words are not always the best medium for communication. A picture can often impart information instantaneously in a case which would require skilful writing and a lengthy time to read.

Engineers have primarily used drawings to communicate design information. But sometimes comprehension of a drawing presents difficulties and care has to be taken in the choice of projection used. The combination of an audio and video presentation often supplies the ideal, a concept which has secured the popularity of television.

1.4.2 Man-to-machine communication

Often communication is only thought of as being directed from person to person, especially as a two-way communication system which greatly aids comprehension. But it is also possible for a person to communicate with a machine and a machine with another machine. Communication with machines is in some ways even more difficult than between human beings. Today, scientists, chartered engineers and technician engineers will need to understand how to communicate with computers in order to improve their efficiency.

The majority of engineers tend to be problem-orientated rather than machine-orientated and speaking to a computer is more difficult than simply translating human statements into machine actions. The machine takes everything literally while human beings use subjective experience to compensate for language inadequacies. This is why, as mentioned previously, the analogy between human communication and inanimate communication systems cannot take us very far: the difficulty being that an electronic engineer can construct two computers with the sure knowledge that they will not fall out, or fall in love with each other. With human beings you cannot be so sure!

The difference is further exemplified by the fact that the human brain can have an inexact memory system whereas a computer has a dull but exact memory. A human being picks and chooses and modifies information received and consequently what is recalled is unlikely to be the same as that received. Also a computer cannot laugh, an attribute which is required for creativity it seems [5].

The computing world recognises two types of language—firstly, the machine-orientated and then the problem-orientated. Programming a computer means writing a sequence of instructions that will accomplish a desired task on a machine. For engineers to talk to digital computers requires that they have a clear understanding of operating procedures. This means that a considerable amount of pre-planning has to go into even the simplest work to be completed before full-scale processing proceeds. This has had its advantages, for it requires engineers to think logically about their problems and rationalise routines, so ensuring that nothing is omitted and that realistic solutions are obtained.

1.4.3 Machine languages

The complexity of communicating with a computer stems from the fact that it comprises a large number of very simple units, not a moderate number of more elaborate units. These basic simple units are electronic or magnetic devices which can recognise two states only, on and off.

Consequently anything that can be communicated to the computer must be expressed in the form of a series of symbols of only two types denoted by 0 and 1. This is no new situation. Before the advent of the telephone, messages were conveyed by wire or radio in morse code (dots and dashes). Since 26 letters and 10 numerals can be expressed by combinations of these dots and dashes or 0s and 1s, anything that can be written in English can be translated in terms of 0s and 1s or binary form (see Appendix 2 for further details of the 'bit' or binary digit).

To give a comprehensive account of various computer languages as they now exist would be like trying to write an account of the spoken languages and dialects of the world. Basically it may be said that there are three classes of languages of increasing order of human convenience.

Class 1 language is 'machine' language which is the binary form of the computer's instruction code—the set of instructions that set in motion the different operations of which the machine is capable. Programming a piece of work for a machine involves breaking the job down into units each of which corresponds to a computer operation, and writing down for each unit the instruction in machine language. The result is a list of blocks of numerically coded information ready for acceptance by the sensory mechanism of the machine.

Class 2 language is an 'assembly' language and stems from the fact that most machine programmes use only a limited number of blocks of instructions. If, therefore, each block is assembled in the permanent memory and given a name, the programmer need only specify the block name and the machine will accept this as a block instruction and interpret it in detailed machine language. Assembly languages are usually expressed in mnemonic form to help the programmer, usually in ordinary alphabetic form with words of six or fewer letters, e.g. SUMSQ for 'sum of the squares of', or XECA for 'execute the instruction in location A'.

Assembly languages can and do grow in the sense that there are added to them new words, i.e. new instruction blocks. Most of them are confined in application to particular machines, but there are some very important exceptions. An example of an assembly language is ICL PLAN used on ICL 1900 series computers.

Class 3 language is called 'compiler' language; this is more general than the assembly language in the sense that it is constructed with the solution of problems in mind, rather than the blocks of operations of which the computer is capable. In a compiler language the program is prepared as a series of written statements—sometimes in mnemonic form and sometimes in a kind of basic English and a previously prepared 'compiler' program takes these statements and translates them into machine language. Four of the most common computer languages are given in Appendix 3.

To give a comprehensive account of various computer languages as they now exist would be like trying to write an account of the spoken the language. It has to be written grammatically correctly and unambiguously.

Even the omission of a point or a misplaced point will give trouble. A typical COBOL listing for part of a run is given in Appendix 4, together with a simple BASIC program for use on a remote computer terminal.

With the advent of interactive computer graphics a new communication link has been achieved between man and machine. Here a language that is familiar to man is used. In place of symbols there are drawings. An engineering drawing is really a universal visual language and engineers have for years been trained to communicate by means of sketches and drawings. The design of a product is realised by transferring information from the designer to the production unit mostly via drawings. Now by means of a light pen (voltage pencil), the relationship between the designer and the computer becomes intimate and is characterised by real-time operation. It is now possible for an engineer to draw on a cathode ray tube and then store, retrieve, modify and rotate designs by means of the computer. The mind boggles at the future possibilities that arise from this conversational mode of communication with the computer. Certainly it may revolutionise design work if it does not prove too expensive.

1.5 Further reading

A Bodmer, F., *The Loom of Language*, George Allen & Unwin, London (1944).

B Hartley, R. V. L., 'Transmission of information', *Bell Systems Technical Journal* (1928).

C Pierce, J. R., *Symbols, Signs and Noise—the Nature and Process of Communication*, Hutchinson, London (1962).

D Gabor, D., 'Theory of communication', *Journal of the Institution of Electrical Engineers*, London, Part III (1946).

E Shannon, C. E., 'Communication in the presence of noise', *Proceedings of the Institution of Radio Engineers*, **37** (1949).

F Shannon, C. E. and Weaver, W., 'The mathematical theory of communication', *Bell Systems Technical Journal*, **27** (1948).

G Cherry, C., *On Human Communication: a review, a survey and a criticism*, MIT Press (1968).

H Parry, J., *The Psychology of Human Communication*, University of London Press (1967).

I Miller, G. A., *The Psychology of Communication*, Seven essays. Pelican (1970).

J Schon, D. A., *Displacement of Concepts*, Tavistock Publications (1963).

K Garbett, J., *The Manager's Responsibility for Communication*, Industrial Society Booklet (1965).

L Evans, I. B., *Studies in Communication*, Secker and Warburg (1955).

See also Reference 1.

2 Communication in the technical realm

Clearly the problem of communicating technological knowledge is one of extreme urgency in this day and age. How can the technologist and technician revivify his specialist knowledge? What should be taught so that technical obsolescence does not come too quickly? How can the learning process be improved? Can technical knowledge be communicated in a more meaningful way? The explosive growth in technical literature is indicated in Fig. 5 (see also Appendix 2). This rapid expansion tends to drive technologists and technicians to specialise and this in turn makes cross or lateral communication between specialists more difficult. There is a dearth of state-of-the-art articles and publications that set out to gather together progress in various related disciplines.

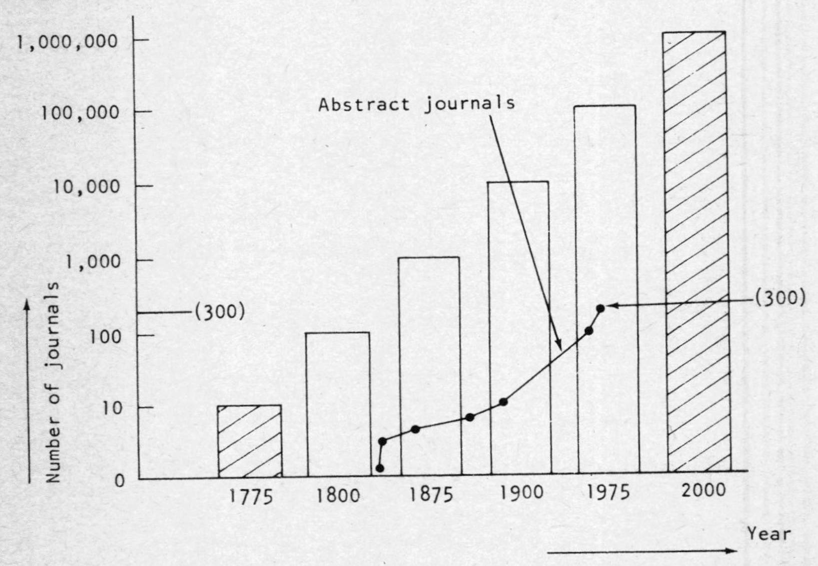

Figure 5 The information explosion in technology. Note: there are more than 60,000 books published annually in engineering and the physical and life sciences. In addition there are approximately 100,000 research reports that remain outside the normal channels of publication and cataloguing

Today professionals in careers based on technical knowledge face the danger of new technologies making their knowledge obsolete and their careers precarious. The pace of change has outstripped man's capacity to comprehend and it is difficult to adjust to new ideas. It is currently estimated that half of what a graduate engineer has been taught will be obsolete in ten years [6, 7]. The recent analysis of 4,000 engineers' career histories has revealed that knowledge obsolescence is of two kinds. There is the loss of once-held knowledge and the failure to learn about new knowledge. The former has been called current knowledge and the latter emergent knowledge. The time dependence of both kinds of knowledge is alarming.

For engineers whose first job fell in the years between 1930–39 there was hardly any obsolescence and there was little difference from one job to another. Engineers working in rapidly developing industries, for example, aircraft, chemicals and scientific instruments, who might have been expected to suffer most showed no more than 9 per cent obsolescence rate in current knowledge, and 14 per cent in emergent knowledge. But by 1964 the picture had completely changed. These same engineers now had a 42 per cent obsolescence rate in current knowledge and 20 per cent obsolescence in emergent knowledge. The same figures for administrators were 17 per cent for current and 40 per cent for emergent knowledge. The problem that obsolescence poses is how a professional career may be sustained unless careful updating is applied.

Hence the need for technical refresher courses and good technical communication of all kinds. Such sources of new information for a professional man form a vital intellectual survival kit, for personal senescence and obsolescence tend to become synonymous.

There is however a further reason for pursuing good technical writing, quite apart from the dissemination of knowledge and experience. Writing and speaking can help to clarify the mind and also assists in an individual's own development. Every technologist needs to recall Bacon's famous aphorism:

> 'Reading maketh a full man,
> Conference a ready man,
> Writing an exact man.'

There is nothing like writing a report to clarify one's thinking and to find out if there are any missing factors. Individuals in both engineering and pure science may derive a sense of achievement and satisfaction when a job has been finally technically documented. But this takes hard work, for writing is like mining not fishing and the satisfaction comes after the document has been completed.

2.1 Communication and technical education

Do we apply scientific analysis to our educational and training communication systems? The answer is generally no. Teaching is communication, but what should be taught and how can the right blend of knowledge and experience be put across? Take the case of a management course for, say, middle managers in an engineering firm. The course might be directed towards the role of finance in business and cover such aspects as understanding the balance sheet and sources of capital, investment appraisal, budgeting and overhead accounting, cash flow, standard costing, marginal income concepts, break-even analysis, discounted cash flow and other quantitative analysis techniques. Depending on presentation these can all be very dull subjects. Care is necessary to ensure that the listeners participate and that apposite illustrations are used. In short, the message must be humanised.

How often does a communicator consider how best to put across his subject matter? Will it be by lecture of the chalk and talk type or by film or slides, programmed instruction, programmed tutorials or project exercises? Could closed-circuit television be used or video-tapes? Far too often dry, dull and dogmatic preaching sessions take place where listeners are forced into a state of lethargy and so become incapable of asking intelligent questions at the end of the session—if any time is left. Yet it is common knowledge that feed-back of reception leads to better learning. 'What I hear I forget, what I see I remember, what I do I understand'. In short, participation is the key factor. Figure 6 shows a case for feeding back results immediately after each test in a simple addition experiment. The same applies to manual skills and Fig. 7 shows results of an aircraft gun turret performance. Programmed instruction, whether in book or machine form does achieve this quick feedback and so applies reinforcement to the learning situation. Too often technical subjects are taught in this country in a manner that leads students into a harbour where they are automatically protected from the realities of everyday life, and the social setting of the problem is ignored. They are rarely presented with information that

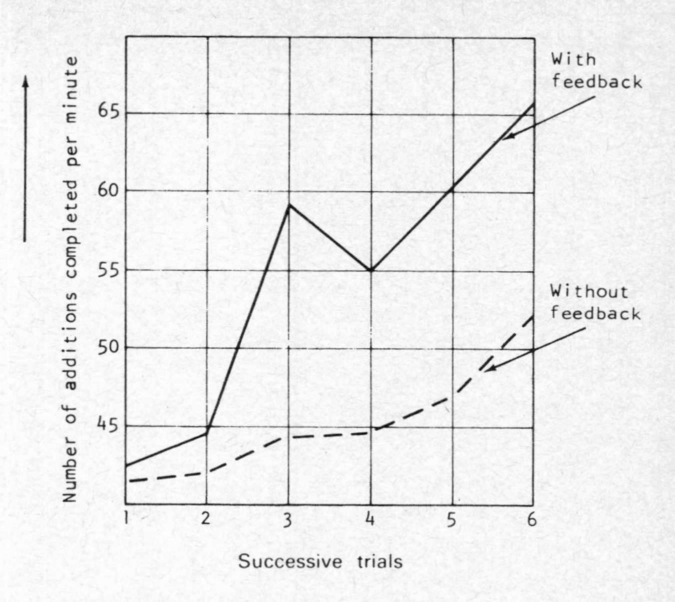

Figure 6 The importance of feedback for mental work

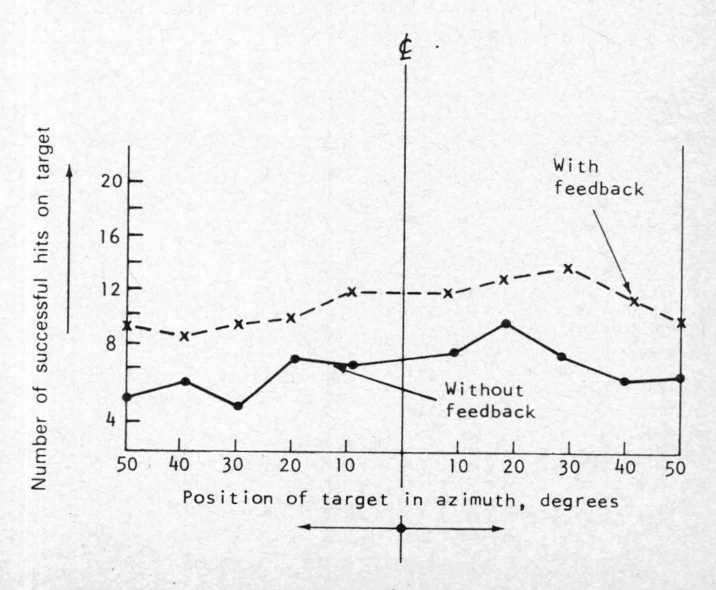

Figure 7 The importance of feedback for manual work

needs a good deal of sifting, the sort of information they will often meet outside academic life. The fact that the lectures are themselves mainly concerned with theory is no excuse for the total disregard of other factors. Most students are forced to work on their own instead of in teams. Money is of no account and the time factor is often uninfluential. In short, a dichotomy exists between academic technical teaching and practical down-to-earth engineering of everyday life which demands accurate technical communication. It is vital for engineers and scientists to be taught in their early days how to learn and where to go to obtain relevant knowledge. Only by close collaboration between academics and employers can this be successfully achieved.

The publication of *Warwick University Limited* [8] demonstrated one instance of the shrinking horror with which many academics regard industry and the business world. It also raised the general issue of the role of universities and polytechnics within contemporary society.

There is no doubt that in the United States the close involvement of university research departments, and particularly those concerned with applied research, has been responsible for considerable speedy technological innovation. The open communication channel has proved to be a source of economic strength and vitality. The formation of Route 128 around Boston has provided a special type of university–associated industrial development forming a necklace of high technology companies spun off and staffed from the academic world. Only by this wholesale transfer of staff has the scheme been made possible. Perhaps in the UK there has been too poor a gearing between the academic world and real competitive business enterprises. As far as engineering is concerned the art of engineering is too often divorced from the science of engineering as indicated in Fig. 8. This may well be due to the fact that the practice of engineering builds from one way of doing things to many and works by synthesis, whereas in the academic world from experimental laws which are obtained by analysis. Obviously what is required is a much freer interchange of people between the two establishments if only to avoid industrial and intellectual authorities being set up. The carry over of technical information and know-how for speedy translation of ideas into saleable and usable products is now a national necessity and can only be achieved by close liaison with adequate technical communication acting as a vital lubricant [D].

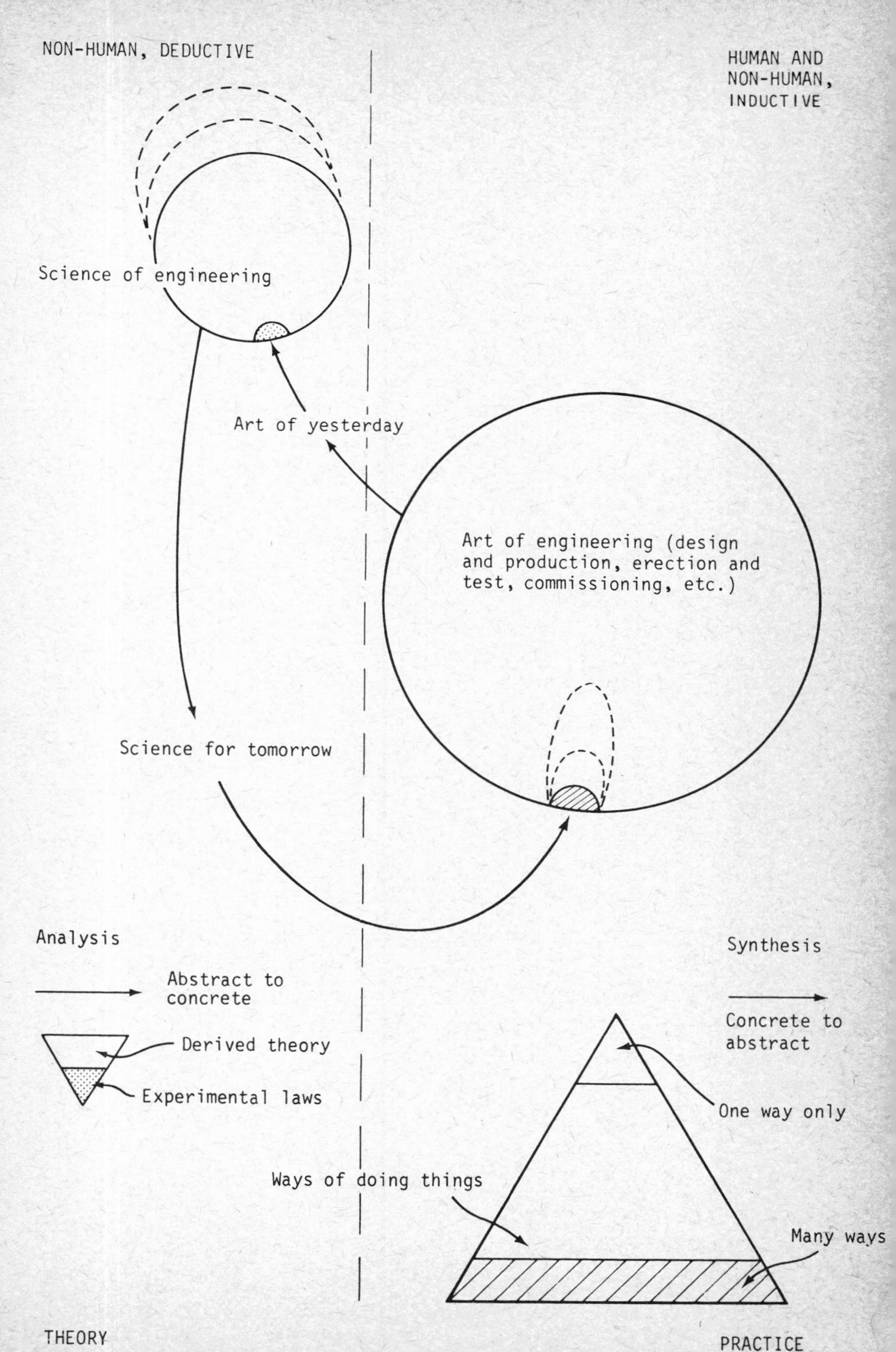

Figure 8 Gearing between the science and the art of engineering

2.2 Communication and language problems

Some say that the reason for much inarticulate technical writing is that the structure of the English language is too rigid and in attempting to force engineering and scientific ideas into it, either the idea or the

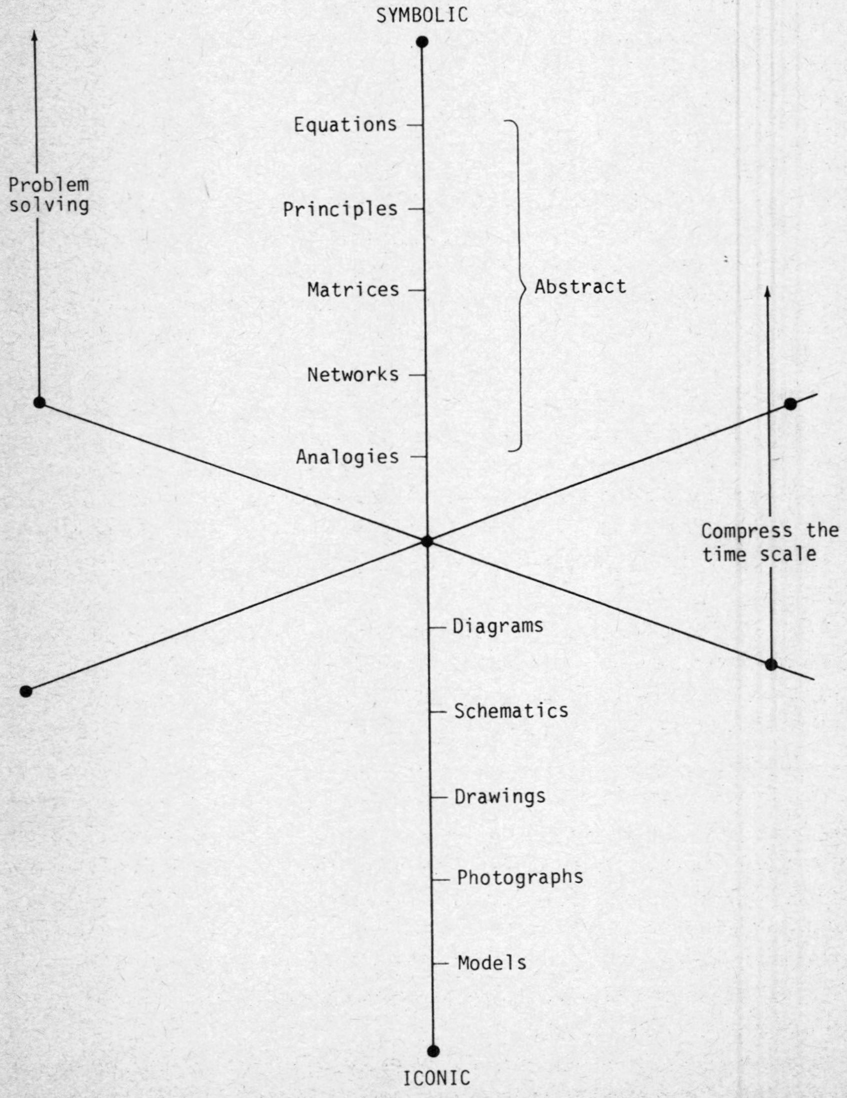

Figure 9 The languages of technical communication

grammar becomes distorted [11]. The types of languages used by scientists and engineers are shown diagrammatically in Figure 9. Here the languages are displayed on a dipole with Symbolic at one end and Iconic at the other. As ideas proceed from the specification stage to the production of hardware so the language of communication used changes from the symbolic to the iconic. In this book we are mainly concerned with written and spoken languages. But in industry continuous translation and retranslation has to take place. As ideas are formulated and manipulated the abstract language of mathematics is used but eventually the ideas are turned into working instructions in the form of diagrams, drawings and models so that production can occur. The symbolic languages are particularly useful for problem-solving while the iconic lend themselves to transferring information on shape and size and special arrangements for manufacture and construction. Table 1 sets out the skills required when using these different languages in proceeding from the mathematical and verbal modes of expression to the graphical.

Literacy in graphical communication is essential to technologists, as engineering design has tended to move further upstream from the drawing boards and thus graphical communication in the form of sketching has become important. A good freehand sketch may be likened to a well written message. The two modes of expression may well need to be used together, a point that will be discussed later in Chapters 6 and 9.

When writing and talking about human situations as opposed to inanimate systems it is vital to put oneself into other people's positions and see and feel what they see and feel. For if there is no empathy there will be no sympathy or sensitivity for the situation being discussed. Said the Prophet 'I sat where they sat' and naturally he could feel for their problems and consequently became a good communicator. The novelist, essayist and journalist are constantly in this position. Possibly one could consider this to be represented by the emphatic/semantic dipole shown in Fig. 10. Here an awareness of other human beings' situations may be heightened by such activities as group analysis and sensitivity training (T groups). To express these meaningfully we need to understand the rituals, rules and rubrics which are set out at the semantic end of the dipole.

A good example of where both the emphatic/semantic and symbolic/iconic have been used together are to be found in Nevil Shute's books

Table 1 Languages and skills required for transmission and reception

Language of communication	Skills required for transmission and reception			
	Literacy	Technical	Non-technical	Competence
Mathematical symbols	Logic Numeracy	Typing	Handwriting	Pattern recognition and significance Generalisation
Word symbols, verbal (written)	Grammar Spelling Structure Vocabulary	Typing	Handwriting	Semantics Syntax Material arrangement Clarity
Word sounds, verbal (oral)	Grammar Structure Vocabulary	Elocution: pause emphasis loudness/ softness intonation	Stance?	Semantics Syntax Intonation
Graphical	Descriptive geometry rules, standards and conventions	Drafting	Sketching	Expressing ideas in 3D on 2D surface

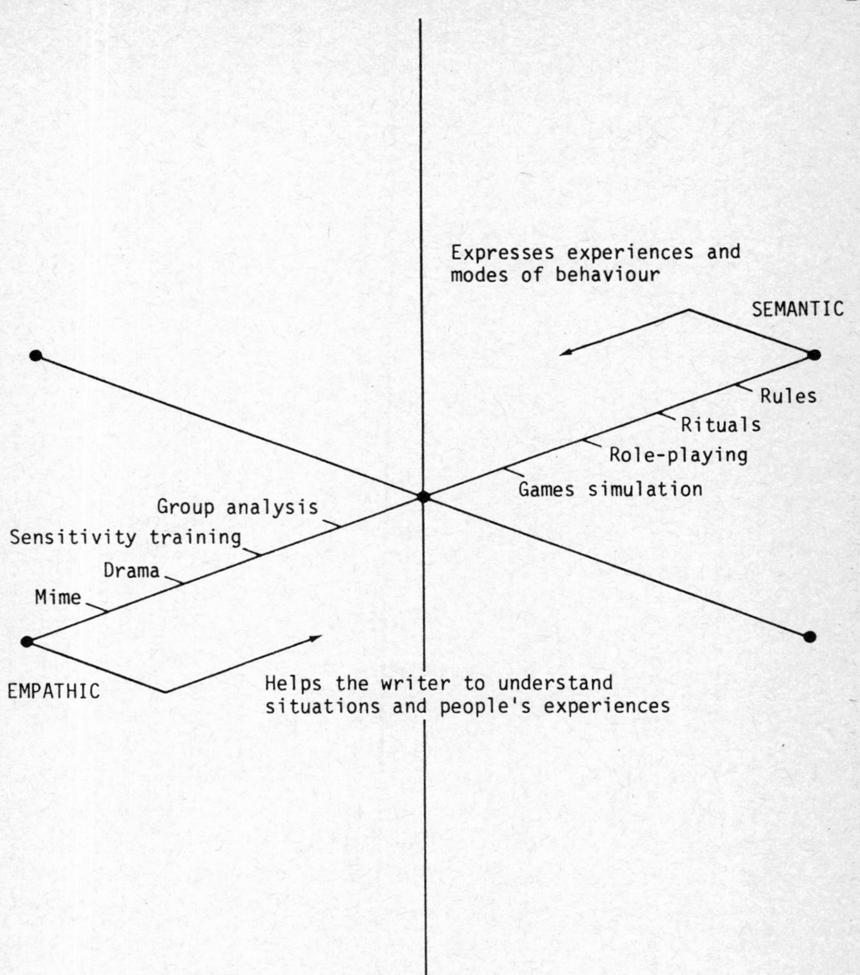

Figure 10 Some factors in general communication: the empathic–semantic dipole

such as *On the Beach*, *No Highway*, *Slide Rule* and *Chequer Board*. Although he hardly ever uses a drawing he often refers to diagrams and models, etc. An even better example is James D. Watson's *The Double Helix*. Here a book tells in narrative fashion a personal account of the discovery of the structure of DNA using photographs, diagrams and figures to back up the text and at the same time gives a sense of the political intrigue, bullying and spying upon competitive efforts which accompanies such work. This kind of writing has something of the 'thriller' about it and knows nothing about the dull grammatical construction so often peculiar to the technical paper—the impersonal passive ('It was earlier shown that . . .'—see Appendix 8). C. P. Snow has used much the same approach in *The Affair* and *The New Men*. But in spite of these efforts somehow one feels that the scientific and engineering world has yet to produce its equivalent of the C. S. Lewis of the theological world. It is not within the scope of this book to discuss such writings but undoubtedly the lay world would welcome something more realistic than much of the current scientific fiction that is produced. To write effectively in this kind of field one must not only have empathy but also an appreciation of the games people play. Inevitably the technologist of today by dint of his education finds it hard to overcome the inadequacy of his equation models of the physical world when extrapolated to real people and world situations.

Generally speaking many engineers have had their enthusiasm for writing blunted by their school experiences, where insufficient attention has been given to writing on practical subjects. The overwhelming emphasis on so-called style, the distressing intricacies of English grammar, to say nothing of some of the spelling hazards have often discouraged many young budding engineers from writing [B].

Somehow there is a widespread attitude amongst technologists of all kinds that communication belongs to the arts and is a lower grade activity compared with design, research and development work. C. P. Snow has already made this clear in his two cultures [9] when talking about engineers and scientists and comparing them with those who have read the arts and has shown that there are faults on both sides.

> '. . . a good many times I have been present at gatherings of people who, by the standards of the traditional culture, are thought highly educated and who have with considerable gusto been expressing their incredulity at the illiteracy of scientists. Once or twice I have been provoked and have asked the company how many of them could describe the second law of thermodynamics. The response was cold: it was also negative. Yet I was asking something which is about the scientific equivalent of: Have you read a work of Shakespeare's.'

Hudson [10, 11] has postulated two types of thinking, each with its own mode of expression. One he has called sequential, and this is used in mathematics, science and engineering. The other, called associative, which is used in history, literature and arts. Some suggest that these are incompatible. Sequential thinking builds up from one premise by deductive logic or *a priori* reasoning. (Because of this—therefore that). This leads to 'convergent' thinking which imposes serious unconscious restraint on the freedom of thought of scientists and engineers. By contrast associative thinking proceeds from one thought to another by various connections—patterns of likeness, spatial relationships or emotional factors. The writer suggests certain relationships (this rather than that) which may or may not be correct. It is essentially 'divergent' thinking leading to a wider view.

It is often said by laymen that scientists and engineers have a language of their own. Special terms are often essential for description but if used should be explained. A spade should be called a spade and a polymorphonuclear leucocyte a polymorphonuclear leucocyte. Apart from mere jargon it is no good purging manuscripts of scientific terminology. This is not the main language difficulty—rather it is in the extraordinary syntax that scientists and engineers use, when writing about their particular speciality or when they get on their feet at a meeting or when appearing on television. Too many engineers move into the passive contorted style the moment they begin to talk or write about lasers or clystrons. With the passive voice comes the inverted sentence. 'Given, as we were, only few data and these not being reliable to two places of decimals, the results which must therefore be taken with a grain of salt and never used to calculate the attenuation, are set out in Table 31.' By the time the reader looks at Table 31 he will have forgotten what it is supposed to contain.

The golden rule is to put the main message in the first sentence and never split it. If there has to be a qualification let it follow in a new sentence. Professor Haldane said that scientists are apt to write their reports of experiments in the third person, saying not 'I observed that' but 'it was observed that' thus sacrificing elegance and paying tribute to objectivity. This frantic effort to be impersonal, so much a part of the scientific method, helps to preserve the artificiality of many scientific papers. Indeed it may well conceal even worse ills such as hiding behind a verbal smoke screen so that outsiders cannot realise that there is nothing unique about the speaker's or writer's findings. A rather good example—perhaps slightly exaggerated—of the style of much technical writing when given as an everyday conversation is shown in Appendix 8.

Probably the grotesque syntax is perpetrated by force of habit taught at school and enforced in higher educational establishments [C].

2.3 Communication barriers

Besides the linguistic barrier and other preoccupation barriers shown in Fig. 3 there seems to be a growing emotional barrier among technologists themselves. The pecking order in the hierarchy appears to be that research is superior to development, development is way above design and production, and so on. Even a highly qualified production manager is looked at askance by a research director who conveys his information in esoteric reports with yards of formulae. All these attitudes lead to watertight compartments being set-up where communication across interfaces becomes difficult.

Perhaps engineers and scientists have themselves exhibited hubris suggesting that they are a superior group. This intellectual arrogance has led them to produce highly specialised polysyllabic jargon frequently used to make truisms sound portentous. Basically society in our country has accorded more status to those who are skilled in languages (particularly 'dead' languages) than to those who are skilled in manipulating numbers and materials. It is very noticeable that participants in management courses who come from the arts side hide their lack of numeracy under a cloud of disinterest and false jocularity. Certainly the position is improving with the efforts of the business schools, for every modern manager must reconcile himself to the need for greater numeracy. Until parity of esteem is afforded to engineering and science there will probably remain some barriers to successful communication of technical problems and achievements to the lay public and even among scientists and engineers themselves. Nevertheless there is little excuse for the use of poor English in so much of our current technical communication.

It is easy to exaggerate by plucking out of context. However, a recent university engineering department report contained the following headings:

> 'Optimisation of non-linear control systems—a simple heuristic explanation of Pontryagin's method based on the use of isochromal hypersurfaces in phase space, has been proposed. Pontryagin's method has been verified in one example by a method involving perturbation of the optimum switching curve.'

So now we know or do we? It depends on who the report was written for. If it was intended for inclusion in the technical notes of the *Royal Aeronautical Society Journal* it might be acceptable as it stands, for Pontryagin's maximum principle is often referred to in connection with space vehicle orbits. (In the March 1972 issue of this journal it was mentioned in three separate articles.) If, however, it was intended for a wider technical public a better heading would have given a simple lead into the subject explaining the wide application of Pontryagin's method. For the lay public the heading is hopeless. It is always a good policy when writing for readers who are less technically knowledgeable to write as if you were taking them into your confidence and try to draw meaningful comparisons with something with which they are familiar. Perhaps a few engineer teachers should be introduced into our secondary school education system in order to inject some sense into writing clearly about practical as opposed to imaginative subjects. One factor is certain: loose or vague writing on technical subjects too often betrays loose and vague thinking about the technical aspects of the subject being discussed. Technical writing can help clarify the writer's idea and often exposes areas of ignorance which previously he had overlooked. You must not just start writing, you must have something to say which you understand and wish to convey to someone else who should also want to understand your message.

2.4 Further reading

A Kapp, R. O., *The Presentation of Technical Information*, Constable (1948).

B Calder, Nigel, 'What they read and why', *Problems of Progress in Industry*, No. 4, HMSO (1959).

C *University Teaching Methods*, Report by University Grants Committee, HMSO, 1964.

D Jenkins, D., *The Educated Society*, Faber and Faber (1966).

E Weinberg, A. M., 'Scientific communication', *International Science and Technology* (April 1963).

F Hutchings, D., *Education for Industry*, Longmans, Green (1968).

G Morris, P., *The Experience of Higher Education*, Routledge and Kegan Paul (1964).

H Turner, B. T., 'Improving communication in design and production', *Journal of the Institution of Production Engineers* (September 1968).

I George, F., 'Barriers to communication', *The Technologist*, **12** (4), (1966).

J Hudson, L., *Frames of Mind, Ability, Perception and Self-Perception in the Arts and Sciences*, Penguin (1970).

See also References 7 and 11.

3 Specifications and standards

Both specifications and standards form an important branch of technical writing. *Specifications* are essentially a statement of requirements to meet a need whereas *standards* are single solutions to repetitive problems. The guiding principle of standardisation is 'the elimination of unnecessary variety'. Codes of practice are less specific than formal standards. They seek to give general guidance rather than detailed instructions, and set out procedures which are currently recognised as good practice. Both specifications and standards are tools of management.

3.1 Specification writing

The main purpose of any specification is to define precisely and completely. All specifications are descriptions in precise terms. Specifications are a part of technical writing, which is a branch of literature, not of technology. The dictionary states that to 'specify' is to prepare a detailed statement of particulars. Documents with this common title serve many, very different, purposes. A specification may be defined as: a document containing a definite, particularised and precise statement of qualities, characteristics and performance of materials, processes and procedures. The statement should be as complete as the state of knowledge and the time and money available will allow.

An appreciation of what makes a good specification in one circumstance is not necessarily helpful in writing for another. There are, therefore, considerable difficulties in writing good specifications and certain qualities are to be found in all these documents but the emphasis and relevance varies with the purpose.

3.1.1 Historical perspective

Specifications began when man first tried to describe to other people the kind of thing he wanted. Some of the earliest specifications are found in the Bible for products such as boats, buildings and food.

One of the oldest specifications extant is dated about 1300 BC and is known as the Eber's papyrus, a prescription which includes measures of quantity of chemical constituents. The Industrial Revolution caused a considerable upsurge in specification writing due to specialisation and the need for consistent quality. The greatly increased knowledge available to man made it possible to define and measure required characteristics more accurately, but also revealed variables hitherto unknown. To obtain repeatability the variables had to be controlled and the degree of control stated, particularly during testing. As the Industrial Revolution changed into the Scientific Revolution complexity of modern equipment and artefacts increased as did the use of mass-produced components. To ensure interchangeability of similar components manufactured at different geographical locations and over long periods, good specifications and standards became essential.

At the same time mankind started to explore new environments and equipment was called upon to withstand higher temperatures and pressures, greatly increased speeds causing various vibration patterns, etc. With new materials to withstand these different environments came the requirement for more reliability, for safety, lower costs of servicing and efficiency of use. As competition grew, higher industrial efficiency was needed to survive, so it became increasingly necessary to avoid over-design but just to meet a specification for a set purpose. Competition has put a top limit as well as a bottom limit on quality.

Defining limits and defining tests to prove limits can be conclusively described as a main function of specifications. As systems grow more and more complex and environments become more onerous, so additional specifications are required to ensure that equipment remains serviceable for a practicable time. This means that better specifications are demanded.

Again increasing complexity means higher equipment cost and problems of manufacturing efficiency increase, causing standards to be written. Remember that a specification is an instrument of standardisation. To standardise is the action taken to regularise a practice. Specification is a much narrower term and means an accurate stipulation of detailed characteristics and other requirements.

Finally, improved control of engineering effort and reduction in waste is required and specifications naturally help to improve these factors.

3.1.2 Types of specification

There are two main types of specification: for things and for actions. Those for things include raw materials, components, final products or systems and complex structures. Those for action include specifications for processes, procedures and performances.

Of course, one document can define both. For example, a component (a thing) may be defined together with how it shall be made (a process) as well as details of tests required (a procedure). All these specifications can state a requirement to be met, i.e. what is the thing or process trying to achieve and under what conditions (i.e. a performance).

A third type of specification is the documents that fall into the category of 'general specifications'. They are guides and contain much relevant information which should be written into each particular specification for equipment in definitive and precise form, e.g. DEF 133.

3.1.3 The evolution of technical specifications

The process required to meet customers' needs can be charted as the successive evolution of documents including specifications. Confusion can result between different types and stages of specification and consistency is one of the quality characteristics of specification (see Section 3.1.4). Many organisations believe that it is necessary to produce a different specification and a different type of specification for each stage of the ideas-to-hardware process. This can be wasteful in time and money. An example follows to illustrate this point.

Take an airborne radar equipment which is required by a customer. First the customer must state his requirements in the form of a user specification. 'We need a radar system to give us information "A" under conditions "B" when carried in aircraft "C".' This may be called the requirement specification. This requirement is now translated into an engineering specification giving precise technical details, such as weight, size, frequency, voltage, maximum temperatures, etc.—a precise statement of what is thought to be possible. There follow further requirements and translations, as a series of black boxes to operate in defined environments for given times, are specified. This becomes a development/design specification. Finally, these black boxes have to be produced and tested and a whole spate of specifications arise: life tests, environment tests and so on. In addition the customer is probably writing a further specification to say how he will prove the complete system when he gets it. The result of this evolutionary growth can be a mountain of paperwork. Each document is probably held by a different person and, worse still, each document may have only incidental similarity to any other.

In fact, every specification should be written first as a statement of requirement then, as more and more is known of this requirement, the specification should be amplified and re-issued. There is then one authoritative 'bible' on the particular product being produced. Where a stable organisation exists, the succession of documents which hand on and amplify the response to the author's need become stereotyped and it is worth determining the form of the specification. It is thus possible to regulate the decision-making process at each stage and to delegate the writing of the specifications as appropriate. It is a particular quirk of engineering history that elaborate rules have been laid down for producing a drawing, e.g. BS308, but very little on the antecedent document—the specification.

When considering the evolution of specifications it is often difficult to differentiate between stating objectives, selecting criteria and specifying. Generally in the design realm, a design brief can be expanded and clarified by designers to set objectives. Selective criteria are the means for setting down objective measurements to satisfy the specification, while specification writing itself is the authoritative way of defining acceptable outcomes for design work.

In starting a design specification it is advisable to try to identify a range of possible solutions and select the widest level of generality to give maximum freedom to the designers. It is also important to realise that at high levels of generality specification writing demands both moral and economic judgements to be made [12].

3.1.4 Essential qualities of good specifications

Among the qualities important in any specification are completeness, relevance, unambiguousness, adequacy and consistency.

Completeness—The purpose of a specification is to define; it must therefore be specific. As far as possible nothing of importance should be omitted or left to the discretion of the reader. The greatest practical difficulty is in matching all the details of the entity with statements.

At the beginning of the translation process, from ideas through instructions to hardware, there can be lack of imagination or communication. In the middle phase of this process, the sheer mass of detail and the task of obtaining correspondence between what is asked for and what is produced is considerable. Likewise, at the end phase: the relations of test conditions to the actual operating environment may be difficult to express accurately.

Relevance—It is useless to specify anything that cannot be achieved by the maker or is already available in a slightly different standard form. (It sometimes happens that 'A' contracts with 'B' to create an idea whose function is already served by an item known to neither of them.) It is also vital that specifications are drawn up in the realm of the possible.

Unambiguousness—The customer may not be too clear about what he really requires. He may merely be expressing wishes or wants that he feels must be satisfied. If so, the writing of a requirement specification is the wrong occasion for clarifying his ideas. Some customers may say, 'I cannot tell you what I want but I will know when I see it'. This nebulous statement does not help the specification writer. Certainly valuable synthesis of ideas can take place before the conception by exchange of sketches and scheme proposals but this activity is but the prelude to specification writing.

Adequacy—Unless the specification is only just sufficient to meet the true needs of the circumstances the cost will be too high in relation to performance. The important word is 'true'. The truth may be very hard to get at, particularly where the customer cannot be consulted directly or where he employs an agent.

Consistency—This has already been mentioned and its importance cannot be overstressed, especially where the final 'bundle' of specifications is used as a control document, as is often the case on American contracts.

A specification often includes several separate requirements which are to be satisfied simultaneously. This is only possible if they are not mutually exclusive. The possibility might not be immediately obvious and if not discovered until late in the contract may need renegotiation.

3.1.5 *Guides to specification writing*

Presentation format and style of layout will necessarily differ from firm to firm, but it is desirable to standardise within any one organisation [13].

Format—A possible format might be as follows:

1 FOREWORD—this must only be used to give background information and the objective of the specification, the authority for its preparation and the name of the issuing authority. It may well highlight particular aspects, such as limitations or revisions in the specification. It must never contain any mandatory requirements.
2 TITLE AND REFERENCE—this should relate to objective.
3 INTRODUCTION—this should seek to place the aim within the context of other objectives.
4 FUNCTION—this must be stated succinctly for either the part or system covered by the aim.
5 FUNCTION—this must be stated succinctly.
6 BOUNDARIES—the use of interfaces, mechanical, electrical thermal, radiation or others which mark off this subject from others.
7 QUANTITATIVE PERFORMANCE—it is best to group together numerical measures in one section so that the reconciliation of separate aims is facilitated.
8 QUALITATIVE PERFORMANCE—wherever possible these should be measurable and put under (7) above. Where this cannot be done, clear, concise statements are needed.
9 REFERENCES—these will be to other specifications, common standards, general information.

Language—It is imperative to adopt rational thinking and precise methods of expression in writing specifications. Clearly a certain degree of language proficiency is necessary and a knowledge of the exact meaning of words.

The language should be terse with the minimum number of words necessary to convey exactly what is intended. Certainly, where possible, figures and other quantitative definitions should be used as they can be more precise than words.

All pseudo-technical words open to subjective definition and judgement such as quiet, light, cheap, typical, smooth finish, instantaneous actions, sufficient strength, as clean as possible, must be avoided like the plague.

It is as well to remember that the words 'should' and 'may' are recommendatory, 'shall' and 'must' are mandatory. There is also a differences between inspect, examine and view.

In order to achieve complete clarity it is often desirable to define terms used. This will include the precise meaning to be applied to words in normal colloquial use, as well as technical terms, if necessary citing a particular dictionary, reference book or standard glossary.

Certainly any symbols and abbreviations used must be given precise meanings to avoid confusion and only standard abbreviations should be used.

General comments—There are several schools of thought on how best to write a specification. One school feels that the designer responsible for that aspect of hardware should also write the specification; another that the customer's engineer should write them too. Yet another feels that though a particular engineer is responsible, his ideas should be put into specification language by a technical clerk. Drafting a specification is a professional activity because it must come between the customer (consumer) and the supplier(s). A document must be produced that supports the interests of both parties, but its prime responsibility is to produce the optimum quality goods to meet the job required with the least effort. Fitness for purpose with value for money. A good specification document will reduce design and development costs and avoid specifying over-stringent conditions that can never be met.

Too often engineers are perfectionists and idealists. They are never satisfied with the adequate and the best becomes the enemy of the good. The specification must control these tendencies. But equally it is found that specifications produced by the customer often overstate the case. The original requirement is sometimes difficult to achieve, i.e. certain details are at times incompatible. Later on the customer feels that there is a possible undefined further use for equipment, so he calls for more stringent conditions than are originally required and feels that this is erring in the correct direction because he will never get what he really asks for. On the other hand the professional specification writer wants to deal only with facts, and if the facts are not available he finds it more useful to state their absence than make up some convenient ones.

In writing a specification one final point needs to be borne in mind and that concerns the need to draw the user's attention to any relevant authorities that should be consulted in relation to the equipment or material to be used, e.g. in manufacture the use of a special process may require clearance of a safety officer before it can be used. This has become particularly important with the Health and Safety at Work Act in the U.K. In general, the writer of any specification should carefully consider what other functions or department could be affected by it and consult them before finalising his draft.

3.2 Writing standards

Basically standards are specifications for items, materials, processes and procedures which are used over and over again. Therefore they must be particularly well written. British Standards, unlike those of other countries, are rarely enforced by law. The few that are, generally concern safety matters. Most are voluntarily prepared and voluntarily applied. Industrial concerns generally set up their own company standards unit which also ensures that International and British Standards are understood and disseminated within the organisation. These units also generally seek to generate their own company standards by collecting relevant data, list it appropriately and eliminating duplication or overlap. At the same time inconsistencies and inadequacies may be revealed which can be rectified by additional standards.

British Standards and International Standards are now produced by extensive committee work conducted by the British Standards Institution and likewise often in-house standards call for committee work [14, 15]. The standards engineer will need to be proficient at running meetings (see Chapter 8).

The whole field of standardisation has come into renewed prominence by the need for metrication. Although the change from one set of units to another may seem a simple matter it is really one of vast complexity since the dimensions and other quantities laid down in standards are intimately related to the units used.

At the same time the re-writing of all specifications in metric terms provides a unique opportunity for a general spring cleaning of British and Company Standards.

3.2.1 Preparing standards

Once a subject has been proposed, data will first be collected which should include relevant national and trade standards. Applications will be noted together with stores inventories and rates of turnover. These will indicate the possible saving that might accrue if a standard were to be produced. A first draft standard will then be prepared for discussion with all those who will be affected by the result. It is at this stage that the technical writing must be good, for it will have to withstand a barrage of questions concerning such things as:

1 Are there too many sizes?
2 Will it cover future development?
3 Is it suitable for all applications?
4 Is it to be applied retrospectively to existing design or only to new design?
5 Is the quality level correct?

and so on. . . .

Only by a very thorough interrogation of all interested parties can a good final standard be drawn up. The process is often tedious and longwinded. The writing consequently is often rather formal but has to be accurate and leave no room for discretion. It will include any relevant specifications and technical data together with inspection requirements such as minimum tests or sampling information. It may also give a list of approved suppliers. Naturally the presentation of standards documents should reflect uniformity of style and clarity of statement and must contain all the information required by any would-be user.

Because standards rigidly defining a product tend to inhibit innovation there has recently been some demand for performance specifications: i.e. those defining acceptance tests and results, rather than production methods, etc. However, these are even more difficult to write adequately.

3.3 Further reading

A Bowles, E. A., 'The preparing of specifications', *Environmental Engineering Quarterly* (October 1963).

B BSI, *Guide to the Preparing of Specifications*, BSI, PD 6112 (May 1967).

C BSI, *The Operation of a Standards Department*, BSI, PD 3542 (December 1959).

D BEMA, *Industrial Control and Electronics Specification Guide*, Publication No. 209 (January 1965).

E EOQC, *International Guide to the Preparation of Specifications.*

See also Reference 12.

4 Contract writing

In the last chapter an agreed and precise specification was considered essential to enable the purchaser and supplier to work satisfactorily together to achieve all the required aspects of performance, working conditions and standards of equipment. In contrast many engineers and technologists think that contract documents are often too inexact. The tortuous and ponderous language is very off-putting and the devious legal wording can repel all but the most steadfast readers. Nevertheless contracts are vitally important legal documents that should lead to the planned utilisation of resources in such a way that both purchaser and supplier receive the maximum benefit.

A contract has been defined as a 'promise or set of promises which the law will enforce'. It is an agreement between competent parties. However a technical contract is also a means of communication between two parties.

The legal elements of a contract are:

1 An intention to create a legal relationship
2 The competence of the parties to do so (sanity, etc.)
3 A lawful subject matter (criminal objectives void a contract)
4 A proper consideration to be given in return for performance of a promise (quid pro quo!)
5 A genuine consent between the parties to do specific things.

4.1 Legal aspects

Today the law of contract is a separate, well defined interest and a substantial body of case law has been built up.

Contracts could be regarded as only one method among many of acquiring a title to property in things personal; others are gifts or grants.

Over the last two hundred years means have multiplied for the creation of complex technological products. Those who devote themselves to the production of this great diversity of material possessions have a considerable advantage over those who buy them. Hence in the interest of society the law interferes with the freedom to make contracts. A good recent example of how the law has intervened has been the various new forms of consumer protection, including the regulation of packaging and labelling of consumer products, the regulation of contracts involving consumer credit, etc.

Technologists are concerned with the scope of contracts in terms of money and time and any special conditions and warranty.

4.2 Scope

The first essential task of an engineer drawing up contract documents is to determine himself the extent of the work to be done and the degree to which the requirements are specific and clear. From consideration of these subjects the contract writer has to decide what form of contract will give his employer the cheapest job in the required time and must fix a contract price that is as definite and as specific as possible [16]. There are really two limitations which every writer must bear in mind:

1 The character of the contract—if it is so complex that both parties must have an army of experts available to draft and respond to its terms it is likely that the written documents will contain so many conflicting objectives as to be unworkable.
2 The risk attached to the contract—subjects for contracts may be ranked in order of certainty of results. In increasing order to technical certainty they concern research, feasibility studies, design, development, production. The financial risk may be in the reverse order.

If the facts and data surrounding the contract are clear, a contract for a specific lump sum will almost certainly be the simplest form. But if certain of the important items that make up the contract are subject to some considerable doubt, or there are other matters which are rather indefinite or where there are difficult construction problems, say in civil engineering contracts, then the engineer may obtain a better deal for his employers by including such items specifically in a separate schedule [A].

The completion date of a contract must be clearly stated in the contract documents and will have an important influence on tender prices. When the time for completion is short, a tendering contractor may have to allow for increased costs of overtime, for payment of premium rates, for additional plant which he might find difficulty in providing at short notice, and so on. Completion on time is not something which just occurs, it has to be planned and worked for. The contract must seek the timely completion by providing an incentive to the contractor and an effective remedy to the employer should there be late completion: usually in the form of a penalty clause—so much rebate for every week of delay.

The contract must state unequivocally what is meant by completion. It could be one of the following.

1 Goods shipped or ready for shipping.
2 Delivery of goods to the purchaser's stores.
3 Physical completion of construction on site.
4 Plant and equipment commissioned and operationally validated.

The type of contract appropriate to a given class of work depends upon the scope of the job, its completion date and the risks incurred. At one end of the spectrum there is the R & D where the novelty of the result being created is high so that there are few appropriate criteria for pricing.

Until fairly recently, Government contracting in the U.K. was on the following basis: actual costs properly and necessarily incurred plus profit at a predetermined percentage. Profit was allowed at 7½ per cent, based on a turnover/capital ratio of 1:1. This is of course completely outdated by recent inflation rates. At the other end of the spectrum there may be lack of cost data as in a production contract. Agreement is generally reached on a fixed price or delayed until the first deliveries are made. In the latter case negotiations are based on drawing information. Profit allowed may vary at 1 or 2 per cent higher in recognition of the additional risk involved to the company accepting a contract on these terms. Where production is continued over a period of time the learning curve technique may be built into the contract. Here the learning curve can be used to establish price year by year. Or the profit arising from any actual improvement in technique may be shared between customer and supplier.

Both the above policies have come in for persistent public and private criticism. The 'cost-plus' method encourages waste and rewards good and bad work alike and generally leads to inefficiency and escalates costs. The 'price to-be-agreed' formula for production work has been shown on occasions to lead to excessive profits, usually because of badly written contracts.

Great efforts have been made to introduce the incentive type of contract with appropriate risks being shared by both parties. In such cases, indeed all types of contracts, the contract writer has to be very clear about how such defined risks are to be allocated and identified (B, C).

4.3 Papers included in the contract documents

The contract documents may include any of the following:

1. Invitation and advertisements to tender.
2. Conditions of tender.
3. General conditions of contract.
4. Special conditions of contract.
5. Specifications of requirement and tender.
6. Plans, drawings and other pictorial representations.
7. Schedule of rates.
8. Formal tender.
9. Correspondence which might take place in the final negotiation stages for the contract, or which accompany the tender.
10. Formal acceptance of tender.
11. Form of bond.
12. Contract agreement.

All of these various items may become binding contract documents and they should never be divorced from one another. Possible exceptions would be the schedule of rates or bill of quantities (in types of contracts in which these are not applicable), the bond (in contracts where that is not applied) and the contract agreement (which is sometimes not called for), etc.

After acceptance of a tender, the complete set of documents must be assembled and in Ministry work in the U.K. it is normal practice to seal them under a formal contract agreement. In industry, however, contracts are not always sealed, but rest on the letter of acceptance. It is desirable that the author of the main contract documents makes an over-all critical review of all the assembled correspondence and separate documents to ensure consistency. It is vital for all concerned to realise that a legally binding contract needs no lawyer or red tape to become enforceable: it need not even be written down, so long as it conforms to the five conditions on page 44, and there are witnesses or evidence of it. So buyers beware.

4.4 Quality of written contract documents

The engineer responsible for preparing the initial documents must always bear in mind that they will be used by contractors in preparing their tenders.

If the scope of the works and services to be performed are clearly and succinctly stated with no ambiguities, this is likely to be reflected in highly competitive and closely spaced tenders, with little price difference between them.

Tendering times are kept as short as seems practicable to the engineer—in some circumstances they can be very 'tight' indeed. Contract documents that are complete, explicit, and logical will enable a contractor's staff quickly to gain a clear appreciation of the nature of the work. Available time may then be used to best advantage, estimating costs with a fairly full knowledge of the project or product. The employer obviously benefits from competent tendering.

Sometimes the cost of preparing the tender is unduly increased—either because adequate investigations of the physical conditions have not been carried out in connection with the proposed work or because the initial documents are ambiguous. If the tenderers spend money in clarifying the situation, the extra cost will eventually come to be charged as higher overheads in the tender prices. If time does not permit the contractor to clarify doubtful points the prudent will increase his tender prices to allow for uncertainties. Others may gamble on being able to challenge the tender document later on and so obtain extras. Hence the great importance of proper wording in such documents. Ambiguity may well be reflected in wide variations between tenders that are received from competing parties as well as in many arguments later on.

4.4.1 Correct wording

The first requirement is clarity of intention in the mind of the drafter. He should practice the use of clear, concise, grammatical English, without colloquialisms, in all contract documents. Too often different words are used for the same item of work in the same documents. Whilst such terms or words may be generally understood, the net result is vagueness; thus disputes arise. A good example of this would be the use of words like 'formation level', 'grade' and 'sub-grade' which are commonly used in earthwork contracts.

Sometimes words are used carelessly in documents, viz. 'water has been purified in the screens'—the water is not pure after it has passed the screens because screens cannot purify water. They can only remove large solid objects.

The verbs 'is' and 'have' are generally overworked. 'The evaporator is to provide a supply of clear water'—'The evaporator has been installed' would have been better. 'The condensers have a vacuum of 29.2″ of mercury'—are working at 29.3″ of mercury.

Always distinguish between 'standard' and 'conventional' and 'outcome' and 'consequence'.

Avoid circumlocutions such as 'in the case of', 'from the point of view', 'in regard to', 'with reference to', 'in this respect', 'due to the fact that'. All such phrases can be used on occasions but constant use leads to sloppy thinking and the use of incorrect nouns and prepositions.

In general, clearer communication is obtained by transmitting meaning through verbs. Adjectives tend to clutter clauses and slow down the reading rate. Basically writers of contracts documents should endeavour to use words of substance and avoid adding vague adjectives to bolster ambiguous words, such as 'plans', 'preparations' and 'situations'. (One often sees 'particular situations', and 'preliminary preparations'.) Further examples of empty words will be found in Appendix 6.

4.4.2 Correct drawings

The contract plans and the specification should give a complete description of the work to be done and of the quality of workmanship required.

Usually, the first documents to be prepared are the contract plans which are, in fact, the engineering drawings developed by the design authority. These must be complete, clear and logical. They represent a pictorial specification and should be free from ambiguity and errors. The drawings must be clearly reproduced and should be logically arranged without undue cramping of details. Sheets should be numbered in logical sequence and indexed. Wherever possible, scales should be uniform and standard for similar items. 'Trades' should not be mixed together on one sheet, viz. structural steel should be on a separate sheet from reinforced concrete work, fabrication on a separate sheet from machining, etc. The aim of the contract writer should be to prepare a planned presentation that builds up a story of what is required.

Typically the first sheet should show the locality with access points to the site clearly marked and be followed by the site plan and general arrangement drawings. Interpretation will be greatly assisted if the orientation of the drawings from sheet to sheet is consistent. In mechanical engineering drawings third-angle projection should be used as this aids ease of comprehension. Where necessary perspective sketches should be introduced to clarify understanding. In some situations contract plans may be indicative only and may not be in final detail. Contracts are sometimes 'let' on such plans and then detailed plans are prepared as the contract proceeds, each one being sufficient only for its particular construction phase. This may enable considerable time to be saved through an early start being made on a project. If this is done, great care should be taken to obtain an accurate bill of quantities with unit prices being specified.

4.4.3 Bill of quantities

The contract schedule, or bill of quantities, or material parts lists will be prepared from the contract plans. Such bills are often the basis of contract payments and, therefore, must give a clear concise financial statement of cost of the work to be performed item by item.

Each item must be defined simply and clearly so that the meaning and scope are definite and precise. This is usually possible if care is taken, but it is sometimes the cause of difficulty. A good example of such a difficulty concerns excavation work. In these days of powerful machinery it is often possible, in the case of open excavation or cutting, to remove soft rock with a suitable machine. This is a cheap method of excavation when compared with the use of drilling and explosives. However, when the rock becomes too hard for 'rooting', then excavation by explosives is essential. In one contract the contractor incurred severe losses from the customer's insistence on the definition that 'rock' was material that could not be 'rooted'. The contractor stated in evidence during the subsequent arbitration hearing that the engineer declined to classify material as rock so long as it was possible to 'root' it—even though it was so tough and hard that it entailed very severe maintenance costs for his earth-moving machinery. For example, he brought evidence to show that he had fractured two gear boxes driving the cutters to execute this work. Here is a case where more careful definition is necessary and it is also clear, where meaning is indicative rather than precise, that there should be reasonable and strictly fair interpretation by the engineer.

Units of measurement must also be explicitly stated, if possible, in terms of an accepted standard measurement. If there is any departure from general practice attention should be drawn to this. Where they affect the fundamentals such as in metric screw threads and other standards, this must be considered from the outset.

A further problem with bills of quantities concerns variations that occur when items are lumped together. Care should be taken with wording to ensure that fair and equitable rates are fixed.

4.4.4. Effect of Tax and Other Legislation

Sales tax may be payable in respect of a contract at both the federal and provincial or state level and may arise at each stage of the chain from ideas to hardware and distribution. The responsibility for collecting tax and remitting same to the appropriate governmental authority may lie with various parties to the contract. Care must be taken to identify whenever taxable transactions are being carried out by taxable people.

Licences or permits may be required before contract work can be commenced or completed and fees may be payable to governmental authorities as a condition precedent to obtaining such licences or permits. All contract writers should be aware of this and responsibility for such fees should be clearly allocated.

In general, government impinges more and more on industry, and there are many areas in which an engineering contract can be affected by legislation. For example, labour laws and industrial safety statutes impose various liabilities on employers and persons responsible in the workplace.

The contract writer must therefore either have up-to-date legal knowledge himself or write in close consultation with his company's legal experts.

A typical example illustrating how contract provisions may assist a contracting party in complying with health and safety legislation would be the inclusion, on behalf of an enterprise purchasing equipment, into their form of general conditions applying to all purchase orders, of the following additional clauses to cover design quality and validation:

(i) The supplier is to ensure as far as it is reasonably practical that the equipment provided under this contract is designed and constructed as to be safe and without risk to health when properly used. Additionally, the supplier is to carry out such tests as are necessary to satisfy the foregoing and to give particulars of testing and examination which has been carried out.

(ii) In order that the purchaser may comply with the provisions of legislation relating to employment standards and industrial safety, the supplier is to provide information regarding any conditions necessary to ensure that when put to use this equipment shall be safe and without risk to health.

The inclusion of these or similar clauses will clearly deal with the vendor's liability and impress on the vendor itself the necessity of becoming familiar with health and safety legislation. It is almost a trite statement to say that knowledge of such legislation is essential to all engineers, whether they practice as consultants, or as employees, and it is particularly important if they are engineering designers.

5 Technical reports and article writing

There is one golden rule to follow in becoming a good technical writer—write simply. This facility is often unnoticed but allows a writer to treat large and difficult concepts without losing readers. As with all art, high achievement is obtained by applying strict discipline. Above all learn the discipline of economy for this will help to win readers. 'Easy writing's vile hard reading', wrote Sheridan, but the reverse is equally true: 'easy reading is difficult to write'.

Good examples of such discipline being applied in technical writing are often to be found in *The New Scientist*, *The Scientific American* and the science and technical notes pages of the *Financial Times*. No Prime Minister in our history has possessed greater mastery over the written word than Sir Winston Churchill. Compared with him Gladstone was incoherent, Disraeli ornate, Salisbury waspish, Asquith prosy and Lloyd George windy!

Note the simplicity of the facsimile produced in Fig. 11 of part of the galley proof of Churchill's *History of the English-speaking peoples*—his eye was ever on the look out for simplicity of expression rather than

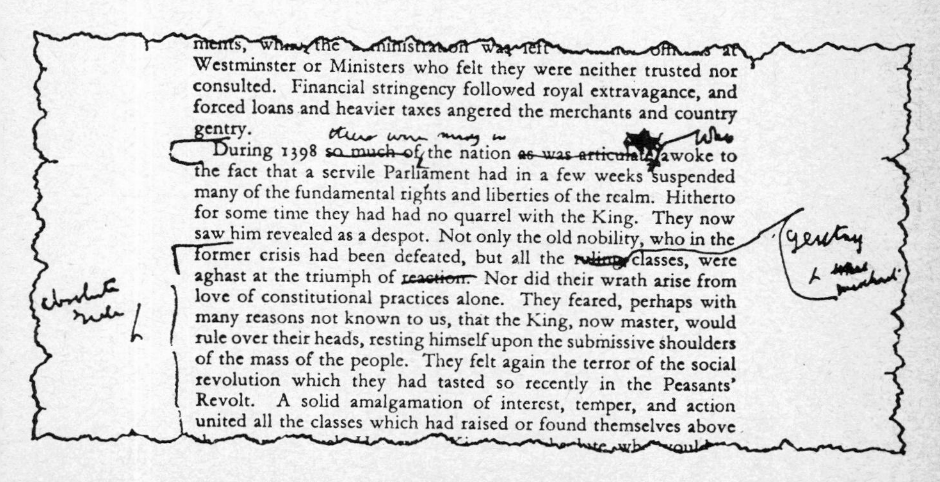

Westminster or Ministers who felt they were neither trusted nor consulted. Financial stringency followed royal extravagance, and forced loans and heavier taxes angered the merchants and country gentry.

During 1398 ~~so much of~~ the nation ~~as was articulate~~ awoke to the fact that a servile Parliament had in a few weeks suspended many of the fundamental rights and liberties of the realm. Hitherto for some time they had had no quarrel with the King. They now saw him revealed as a despot. Not only the old nobility, who in the former crisis had been defeated, but all the ~~ruling~~ classes, were aghast at the triumph of ~~reaction~~. Nor did their wrath arise from love of constitutional practices alone. They feared, perhaps with many reasons not known to us, that the King, now master, would rule over their heads, resting himself upon the submissive shoulders of the mass of the people. They felt again the terror of the social revolution which they had tasted so recently in the Peasants' Revolt. A solid amalgamation of interest, temper, and action united all the classes which had raised or found themselves above

Figure 11 Some editing by Churchill of his own writing

rhetoric. Here 'So much of the nation as was articulate awoke' becomes 'There were many in the nation who awoke'.

One of the reasons why successful technical writing (and all writing for that matter) is difficult to achieve is because it calls for many decisions to be made simultaneously. As a writer one has to decide such matters as:

1 What subject matter to deal with first and what later.
2 What words to use from the rich variety of the English vocabulary.
3 How to form sentences.
4 Which ideas should be emphasised.
5 What should be omitted or referred to elsewhere, etc., etc.

The writing process is arduous and tiring but it is creative. It is only after having written that satisfaction is achieved. Because it is difficult, engineers and scientists tend to adhere to set forms and well worn phrases rather than developing a fresh and readable manner. The purpose of all technical writing is to present information in a good readable style. It must make the recipient want to read and digest what has been written. All authors have, as their first duty, a responsibility to their readers who should be able to follow the text without difficulty, understanding each part as it comes along. This can only be done if the author has first clarified the subject in his own mind. He must decide for what purpose he is writing. Is he trying to introduce a new idea, or report findings or recommend a course of action etc.

Before starting to write it is essential to have all your data assembled, and your thoughts about what you are going to say, clear. At this point you must decide the type of report or journal you are aiming at, and the technical level that can be expected of its readers. If writing a technical article for a journal take the precaution of examining a few back copies for these points: style, presentation and technical level. Remember that every publisher must be a seller. When an editor looks at an author's manuscript he looks further than the written word to the ultimate display of the article or report in his journal. Sometimes it is useful to have a talk with the editor before embarking on writing an article. Having chosen a particular technical level for your work, keep to that level, and don't be tempted to appeal to a different class of reader. If you do, you will risk falling between two stools.

5.1 An orderly presentation—the structure

Having assembled the information you need, your next task is to analyse and organise the presentation of the information into a logical sequence. Wherever possible try to give a preview of your subject as a sort of schematic—and if you can give an actual block diagram so much the better. In any event the labels you assign to each of your blocks, whether verbal or drawn, can be re-introduced as sections in the body of the article. There is no hard and fast rule for this, but a description of some experimental research for example, might be presented under the following headings:—introduction, summary, apparatus, tests, results, discussion, conclusions. When David Fishlock, Science Editor of the *Financial Times* writes on technology and the political, commercial and social consequences of its progress he does so with great clarity because he structures his material so well. In his book *Man Modified* [18] which deals excellently with man/machine interaction he presents his material, after a brief introduction, in six chapters with the headings 'Man Measured', 'Amplified', 'Augmented', 'Mimicked', 'Transplanted' and 'Modified'. Enough to whet the appetite of any technical or non-technical reader.

The purpose of the introduction is to put the reader into the picture, presenting the background to the work, explaining how the need for the research or study arose, what is was hoped to achieve, etc. The section on 'Apparatus' is a simple straightforward description of the apparatus used, including an account of any specially constructed equipment. The section on 'Tests' should contain sufficient information for the reader to follow their purpose and, if necessary, to repeat the tests himself. Any difficulties or problems which arose should be described if relevant to the subject.

'Results and Discussion' could be presented separately or together, and the choice would vary from paper to paper. In presenting your results, however, you should not overlook what you said your intentions were in the introduction. Indeed, the results presented and the discussion should flow naturally from the purpose outlined in the introduction. It is also a good idea to present the results and the discussion in such a way as to bring out more clearly points which you would then gather together and present as 'Conclusions'.

There is a more scientific way of analysing the subject of your report or article based on a principle sometimes used by those who write programmed learning texts. Basically it consists of a matrix where the main headings chosen for the report are numbered and put across the diagonal of the matrix. Each heading is examined for a direct relationship with every other heading. When such a relationship exists the squares that form intersections between two or more related headings are shaded. A typical example of such a matrix is shown in Fig. 12.

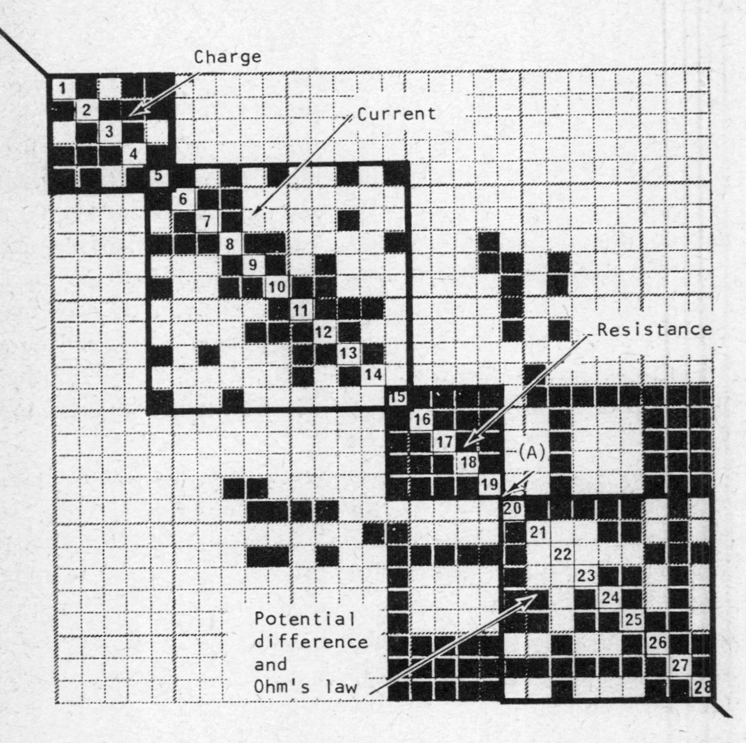

Figure 12 Structure matrix to indicate concept areas and their inter-relationships

An inspection of the completed matrix leads to an easy identification of concept areas and shows how these overlap, butt together or leave a gap. In the figure, which represents part of a matrix for writing about Ohm's law, four concept areas are identified—charge, current flow, resistance and potential difference and Ohm's law itself. Squares 1 to 5 on the diagonal cover the concept of charge, 6–15 current flow, and so on. Thus square 1 has the heading 'Charge produced by friction', square 2 the heading 'An electron', while squares 3 and 4 cover 'Electron as an impractical unit of charge' and 'The coulomb as the practical unit of charge'. Finally square 5's heading becomes 'Current is the flow of charge'. Since the matrix is symmetrical only one half need be drawn in practice. Where breaks occur along the main theme line (the diagonal), as at (A), it suggests that restructuring is necessary to lead the reader gently from concept to concept (in this case from the concept of resistance in Ohm's law to potential difference). For a fuller explanation of the use of this method for report writing see Reference 19. A detailed example is also given in Whitehouse's book *Documentation* [G].

5.2 Actual writing

Having got an overall picture of the article or report in your mind, you can make a start with the actual writing. Much depends upon the type of report you are going to write, but the creative process of writing follows a cyclic pattern (see Fig. 13). The cyclical process starts with a need that sets the objective. Why are you writing this report or article? Is it to inform, persuade, teach, etc. It will determine the target population of the readers. This in turn will determine the method of selecting and presenting ideas and information.

Once the above are clearly defined the next stage is to collect the relevant facts, data and other information and reference material. This will then need to be analysed and placed into an appropriate structure, at the same time making sure that there is a logical sequence as previously discussed.

Only when these steps have been completed can you plan the writing and start to write. You do not necessarily have to write from the 'Introduction' to the 'Conclusion' consecutively, you might wish to start by writing the chapter on 'Experiments performed', perhaps because these are fresh in your mind, or are complex to set down in words and you wish to get the difficult part done first. When all sections have been written then they may be fitted together. One further point is the importance of having a time scale to work to. Most writers work best when the time scale is tight but you need to remember the last part of the cyclic process and allow time for rewriting parts, or whole sections, editing and polishing. The stages from the first draft to the final polished copy are iterative and the aspects to look at are set out in Table 2. (Another way of expressing this phase is given in Fig. 27.)

In choosing the title of your report be careful to incorporate key

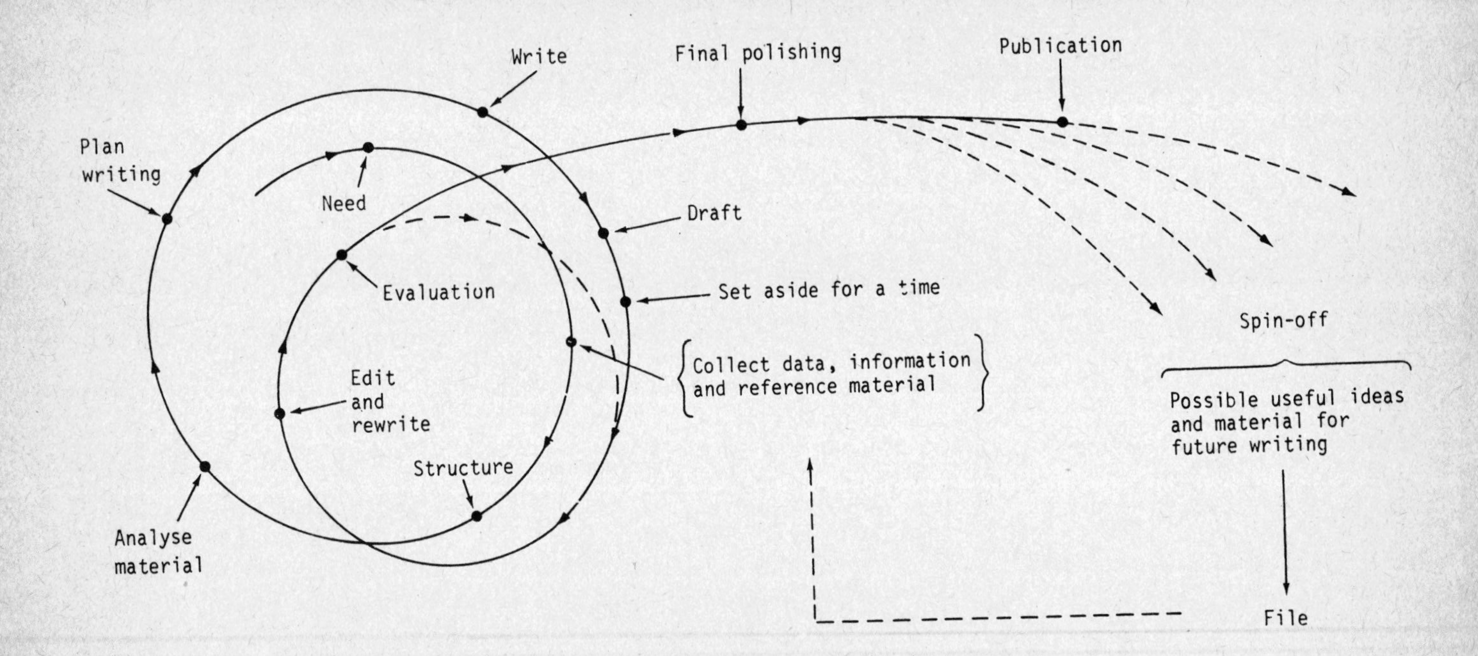

Figure 13 The creative writing process

words wherever possible for this will help information retrieval.

In any language, some words are less important than others. The total number of words varies according to the level of technology in a country: the English language contains over 50,000 words in general use. But not all these words are used for communication in a particular industry.

Table 2 The iterative phase of the report writing process

Stage	Aspect of review	Check for
Draft	Structural logic	Subjects out of order Duplications Irrelevant material Missing material
Edit and rewrite	Clarity and accuracy	Jargon deletions Grammar (tense) Data errors Maths errors Technical conclusions Ambiguities
Evaluation	overall style	Abbreviation errors Caption errors Appendix numbering Illustration fit
Final polishing	overall excellence	Readability Consistency of figures and tables, etc.

Probably only 10,000 are relevant and of these not all are essential.

By careful selection it is possible to arrive at a list of essential reference words for a particular business or industry. However, selecting these key words is quite a complicated job and for particular technologies needs to be undertaken by an expert librarian. Where such lists exist writers should choose report titles containing key words or state them separately for easier filing and retrieval.

For example an article entitled *A new versatile microfilm printer* might actually mention that for information retrieval the key words for this article are: Drafting/Microfilm/Printer.

It may be best to choose your title after writing. If necessary, you can incorporate a sub-title to make it more meaningful such as *Manned satellites—their achievement and potential.* After the title you should give your name (complete with professional qualifications, etc., and status in the company). You should also provide a summary of 100–200 words. The summary should contain the essential information, and be capable of separate publication, if required, as an abstract of the article.

Now comes the body of the article. Remember always that your duty is to your readers. Have compassion for them in their ignorance. You are probably ten times better informed on your subject than 99 per cent of them. Before starting the introductory remarks pause to make a solemn vow that you will lead them gently up the path of knowledge, neither over-estimating their familiarity with the subject nor plunging into it too precipitately.

As a result of the thought you have already given to your article, it will be divided into several major sections. Some of these sections should perhaps be broken into sub-sections, and so on. Remember that intelligently used sub-headings are a help to the reader. They give him guidance (and a refreshing break), and in an unillustrated article they add interest to otherwise unrelieved areas of solid type.

A good way to ensure that appropriate headings and sub-headings are chosen is to use the decimal system as used in this book. This gives headings and sub-headings arranged in a hierarchy. For a report on a space vehicle the headings and the numbers might be as follows:

- 1 Introduction
- 2 Space vehicle development programme
- 3 Main technical features
    - 3.1 Airframe
        - 3.1.1 Special structural features
        - 3.1.2 Structural attachments
    - 3.2 Propulsion
        - 3.2.1 Solid motor version
        - 3.2.2 Liquid motor version
            - 3.2.2.1 Pressurised loading
            - 3.2.2.2 Gravity-feed loading
    - 3.3 Plastic components
    - 3.4 'Jinker' and control systems

3.5 Guidance system
    3.5.1 CW system
    3.5.2 Pulsed system
        3.5.2.1 ...

and so on.

This system has the great advantage that extra paragraphs of new orders may be added without renumbering the whole report. Thus if further details on 'special sub-structural features' are deemed necessary after re-reading a new paragraph may be added numbered 3.1.1.1. Cross-reference to specific sections is facilitated too.

5.3 Don't blind 'em with science

An article that cannot be understood by the people for whom it is written is a badly written article. To avoid this fault you should:

1 Write as simply and clearly as you can.
2 Use the shortest possible words in the shortest possible sentences.
3 Avoid foreign words and phrases, Latin and technical jargon that is not universally understood.
4 At all times seek to be clear, precise, logical and brief. In other words, if you can get close to conversational speech in your writing you will quickly achieve a readable and interesting style. Remember what was said in Chapter 2 and look again at Appendix 5.
5 In spite of contrary advice from our mentors over the years there is no doubt that personal pronouns lend vigour and authority to technical writing. They also encourage the use of active (as distinct from passive) verbs, 'It was decided' is vague. 'I decided' is positive and clear.
6 If a colleague were to ask you how your work is going, you wouldn't reel off mathematical formulae at him. Equally in writing it is far better to put your observations and results into words, and then if you feel they need support from mathematics you can add an Appendix at the end. (See Section 9.3.1 for further details regarding mathematics.)

5.4 Verbiage and vagueness

Your writing will never suffer from weakness or vagueness if you keep to concrete (as distinct from abstract) nouns and active verbs. There is a temptation sometimes to use slangy expressions, placed in inverted commas, to lessen formality. The intention is good but the practice should be avoided. It puts up a sign 'Here is a lazy mind that can't be bothered to find the right word'. Perhaps the heading to Section 5.3 is a case in point.

Some people get all worked up about clichés, but there really is no need. Well-known phrases can help both presentation and understanding. Just make sure that you are not using woolly expressions and phrases to hide woolly thoughts. Woolly words abound too; words like adequate, or appropriate. Your reader's ideas of their meaning may be different from your own, so for his sake, be more precise. (There now I've done it! Go back and take out the word 'more'. That's better.)

Other vague words you should think twice about using are 'blueprint' (who uses blueprints these days anyhow?), 'breakdown', 'over-all', 'target'. Beware too of abbreviations comprising the initial letters of several words (acronyms). If you must use them, quote the title in full the first time (with the abbreviation in brackets) and then you will know that all your readers understand.

Sometimes you may find it useful to form an acronym if you intend to keep repeating the same words throughout the article. Thus Semler in his article entitled *What's wrong with the technical press* [20] uses this artifice like this. 'The economics of paper, print and postage (henceforth known as PPP) are such that the. . . .' He thus avoids having to repeat 'paper, print and postage' every time he refers to this point in the remaining part of his article. Value and quantities may also be abbreviated, but stick to the abbreviations recommended in BS500 and BS1991.

5.5 Toppling a few false idols

As English scholars have taught you what grammar you know, and English scholars are by definition purists, your style is probably cramped by meaningless restrictions. 'It is well to remember', wrote Somerset Maugham, 'that grammar is common speech formulated—usage is the only test'. He went on to say that he would prefer a phrase that was easy and unaffected to a phrase that was grammatical. So a half-page devoted to toppling a few false idols won't come amiss.

There is no sacred bond between the two parts of the infinitive tense of our verbs, but that is not to say you can split them as you please. Do it only where the sentence reads more easily as a result.

An English scholar may be thrilled by a grammatically correct 200-word sentence, but your reader will be unimpressed and often bemused. Here is a 133-word masterpiece purporting to tell in one sentence how dentists work out their salaries. This was actually published by the National Health Service:

> 'In any succeeding month in the same year the remuneration shall not exceed such sum as will when added to the remuneration of the previous months of the year amount to the product of the standard sum multiplied by the number of months of the year which will have expired at the end of the month in respect of which the calculation is being made together with one-half of any authorised fees in excess of that product which but for the provisions of this regulation would have been payable in those months were you to exclude for all purposes the limitations set out in paragraph 2 of section three of the amended regulations of the month of January 1949.'

No wonder dentists get higher salaries than engineers and scientists!

It's remarkable that a recent Government report on cabbages contained 1,756 words; the Lord's Prayer contains 69 words and for such brevity we can all say 'Amen'. So restrict the length of a sentence to not more than 35 words, unless it is broken up by semi-colons, etc., into two or more distinct and logical parts. A side product of the trend to shorter sentences is that it is now quite common to start off with a conjunction. 'And' and 'but' are the more usual, but others are possible. This can be overdone, for the sheer joy of doing it, but used sensibly it gives strength and impact to writing.

At the other end of the sentence you have probably been warned never to finish with a preposition. But why not? If the words scan best with a preposition at the end by all means put it there. When someone at the Admiralty 'corrected' an officer's report by taking a preposition from the end of a sentence and putting it in the middle, Sir Winston Churchill scribbled his famous marginal comment: 'This is the sort of arrant pedantry up with which I will not put!'

5.6 Words to watch

Much has already been said about words. You are measured by your words. Study how to use words properly. Two commonly misused

English words that frequently appear in technical writing are 'infer' and 'comprise'. To infer is to deduce, not to imply, and to comprise is to consist of, not contain. Here is a short list of other commonly misused words with their meanings [21, 22]:

> To anticipate means foresee and to deal with an event in advance.
> To transpire means to become known.
> Practical means concerned with practice.
> Practicable means that it can be done.
> Alternate (adj) means every other one.
> Alternative means a substitute.

You might like to try your word power by referring to Appendix 12. Since words are intended to communicate information, it is important to choose words that do just that, as quickly, simply and unequivocally as possible. This does not mean that the words must be long and the sentences involved. Quite the contrary. Many years ago a traction engine got out of control and finished up with its smokebox embedded in a public house. A telegram was sent to the engine makers, giving the name and place and reading, 'Engine in pub. Send fitter. Urgent.' That would not have won a prize for English composition, but it was clear and left no doubt in the reader's mind.

Avoid empty words and phrases and beware of using heavy connectives like 'accordingly', 'consequently' and 'more specifically' (see Appendices 6 and 12).

It is not enough that we know what we want to say and that the words chosen should have the approval of the dictionary, and be strung together in a sentence which the grammarian would pronounce faultless. The sentence may still fail to convey the precise meaning intended by its writer or speaker. There are many rules for avoiding this type of confusion; those who wish to follow the matter up can refer to the numerous books on the subject [21, 22].

Long-windedness is a common mistake. It has been called 'barnacular' and 'pudder', or by the Americans 'gobbledegook'. George Orwell has pointed out that the passage in Ecclesiastes about the race not being to the swift nor the battle to the strong could have been written as: success or failure in competitive activities exhibiting no tendency to be commensurate with innate capacity. Both say the same thing; which is better? Before we laugh at such a long-winded construction, let us think how many times we find similar long-windedness in our own technical reports.

It has been said that the best way to tell a story, or to write a technical report or article, is to start at the beginning, go on until you have finished, then stop. But how often do we find that a description of, say, a machine tool, will start by telling us the horse-power of the motor, go on to describe the gearbox, then dart back (because the writer has only just remembered it) to describe the clutch through which the gearbox is driven? Then after a few more false starts and wanderings down intriguing side lanes, it might tell us what the machine does. If only the writer had thought of starting at the beginning!

5.7 Final draft

Before having the final draft typed, heave it out of sight for a week or two and then try to re-read it with a fresh mind. Write in haste, polish at leisure, is a good principle to follow [23]. If you come to a sentence that trips you up at first, try to imagine how much more difficult it would be for any other reader. Ask yourself how you would explain the sentence to a friend and then substitute this explanation for the original in the manuscript.

Never be afraid to cut and change to get the meaning clear, go on polishing and use the cut and paste method to restructure your manuscript until you are satisfied that you have a meaningful message. Remember that professional editors spend 90 per cent of their time crossing out words in manuscripts and shifting around those that are left. You should now ask a friend to read the article critically for you, because appraising your own work is more difficult than pleasing the boss. As an example of appraising a technical article you might like to refer to the example given in Appendix 10. A critical appraisal like this can often clarify one's writing.

Have the final draft typed on one side of the page only, with double or treble line spacing. At least two copies will be needed: one for you, and one (the top copy) for the journal to whom it will be offered. Remember that some technical and scientific journals have papers refereed before publication. A copy for each referee as well as a copy for the Editor is not only helpful but will also help to get your paper published more quickly.

All figures should be separate from the text, and numbered in the order mentioned, regardless of whether they are line or photographic illustrations. Check that they are all in fact mentioned in the text. Line drawings should be clear, and drawn in black waterproof ink on smooth whiteboard or good quality paper. Lettering and numbering on any diagrams should be upright and of 'UNO' stencil or Letraset type in capital letters, and large enough to stand a 3 to 1 reduction. (Note: if Letraset is used it should be sprayed with a fixative.)

Even if you have fully described the illustrations in the body of the article you should still give them long captions saying exactly what they show, and drawing attention to any interesting points. If you include engineering prints they must be as simple and small as possible, consistent with legible lettering, with all unnecessary details crossed boldly off. Remember that all illustrations have to appear a few inches wide on the page which often means a severe reduction.

Photographic illustrations should be at least half-plate size ($6\frac{1}{2} \times 4\frac{3}{4}$ inches) and amateurish prints are seldom satisfactory. They must be clear black and white glossy prints on hard paper and unmounted. You shouldn't make any sort of mark on the front of a photograph intended for reproduction, but if you wish to draw attention to any part of it you can stick transparent paper across the front and pencil lightly on to this. In any event be careful not to mark or score the photograph in any way. Never use paper clips on photographs without first protecting them with paper wads, and never send them through the post without cardboard stiffeners. Creased and marked photographs are useless for reproduction. Remember that suitable acknowledgement must be made if you use a photograph belonging to someone else. You must also have their permission of course. Further details on illustrations and how to use them to best advantage is given in Chapter 9.

You might need to acknowledge help, information or co-operation received during the preparation of the text including your own company. In any case, you must include a section at the end of the article in which you list all publications from which you have quoted.

5.8 Some further points

Learned societies and professional bodies all provide printed notes for authors of papers, and these give very good guidance on presentation, number of words, figures, etc. Technical journals do not often provide similar guidance to intending authors, but it is safe to say that they discourage articles of more than 4,000 words. Topics of major importance are exceptions, and you may find that editors will accept even 8,000 or 10,000 words on these and will publish them in two separate parts in consecutive issues.

Editors welcome an estimate of the total number of words in your contribution, made as a pencilled note at the top of the first page. All that is necessary is for you to count the words on a typical page and then multiply this by the number of pages.

Having had your manuscript typed in double spacing for this very purpose, do not mind the editor putting blue pencil all over your virgin script. He is trying to help and often sees errors, redundancies and ambiguities which you as the writer can all too easily miss.

5.9 Proof reading

Proof reading is really a specialised profession but by being observant and attentive you can learn to spot errors. Since you as author are most familiar with the subject matter you are naturally invaluable as a proof reader. The system for correcting proofs is very different from the one used to correct your manuscript. On your manuscript the editor, or you as author, will have made changes at the point of correction; hence the necessity for having your manuscript typed in double spacing. The compositor then reads your copy line by line, setting the type as he reads. But once the copy has been set 'in galley' and read by a proof reader, the compositor does not read each line but runs his eye down the margin to detect changes that are necessary.

When the 'galley proof' eventually arrives on your desk try to use the standard 'marks' set out in BS5261: Recommendations for proof corrections and copy preparation. The main 'marks' for proof readers are set out in Appendix 11.

When correcting your proof put your remarks in the right or left hand margin, whichever is nearer to the word corrected. If there are several corrections in a single line, place them in order from left to right, separated by a starting line, e.g. bold/. If the same correction is made several times in the same line with no intervening correction, make your correction once in the margin, followed by an appropriate number of starting lines. Thus if you add 's' to three words in a line, write in the margin 's////'. When you wish to insert words, put a caret (insertion sign) at the point of insertion and write the additional words in the margin. To delete material without substituting anything, cross it out and put a delete sign in the margin. Should you inadvertently delete material that on reflection you feel should be restored, all you need to do is to place a row of dots under the crossed out words and cross off the delete sign in the margin and write 'Stet'.

Should you ever have to add a substantial amount of new material then it should be typed at the bottom of the galley or on a separate sheet attached to the galley. It must be identified by a galley number, indicating clearly where it is required to be inserted. Never cut a galley [24].

Sometimes the first proofs the author sees are after the type has been made up into page—these are known as 'page proofs'. Only where it is absolutely essential should extra matter be added on the page proofs since this can cause extreme complications and high correction costs. For example, if you add five lines plus a formula in the middle of a page, the printer has to transfer the last 6 or 8 lines of that page to the top of the next page and so on. When tables, illustrations and headings intervene it may not be easy to balance pages.

If you can make changes so that the number of characters in any line remains the same so much the better. You have to count each letter, digit, punctuation mark or space between words as a character. Then if you insert words, try to delete words having an equal number of characters so that you come out even within the line or within consecutive lines. The same applies to the total number of lines.

You must read your proof quickly and return it to the publisher as soon as possible and please remember *you should not rewrite your paper or report at the proof stage.*

5.10 Report format

When internal company reports are being prepared it is desirable to adopt a distinct style and keep to it. There are many variations but it is essential to have easily readable typography for the title, index and summary. For scientific and engineering work the summary should be placed on the first page. For commercial and military work, proper security classifications must be clearly indicated. A good report numbering system should also be designed so that easy retrieval can be achieved. A further feature should be the inclusion of a distribution list so that all recipients are known at a later date.

Some typical examples of a good report format can be seen in Appendix 9. Here the title is clearly shown in a box, the name of the author, distribution list, date, references, etc., together with a short summary.

The summary is a very important part of your report and should crystallise in a few words the main purpose and findings of your work. It is sometimes possible to get someone familiar with your field of work to read the report and write a summary for you. This can be very illuminating and may show you how badly you have written your report. You may need to rephrase your helper's summary to set the style in line with your report or article. Remember that the majority of managers will only have time to read the summary. A survey taken of managers in several large engineering enterprises revealed that 87 per cent of the managers who received technical reports read the summaries and only 12 per cent referred to the main body of the texts, see Table 3 (see also Reference 25).

Table 3 Parts of technical reports that managers read

Part of report	Percentage of managers reading part*
Summary	87
Introduction	43
Main body	12
Conclusions	55
Appendices	5

* Sample size, 287

5.11 Punctuation

Perhaps a few remarks about punctuation would not be out of place. Whole books have been written about punctuation and there is still considerable disagreement between grammarians on the subject. Gowers [3] says that taste and common sense are more important than any rules, and that stops should be put in to help the reader understand what has been written, not to please the grammarians.

Fashions are changing in the use of stops. At one time there was an idea in existence that every possible stop should be inserted in scientific writing; it was believed that this in some way added to the precision. But this notion has disappeared. The use of too many stops produces a jerky style, and hinders that smooth pouring of information into the mind of the reader which is so much to be sought after.

Fowler's *Modern English Usage* says that anyone who finds himself putting down several commas close to one another should examine the text to find if any can be omitted; if none can, the sentence should be re-shaped.

He also says: 'Those who are learning to write should make a practice of putting down all they want to say without stops first; what then, on reading over, naturally arranges itself contrary to the intention should should not be punctuated, but altered; . . .'.

Gowers gives an example of the sort of sentence that Fowler has in mind. A colonial bishop wrote to *The Times* a letter containing the sentence: 'I should like to plead with some of those men who now feel ashamed to join the Colonial Service'. The absence of a comma after ashamed makes a world of difference to the meaning! But it is unwise to construct sentences in such a form that so small an omission can so drastically change the meaning.

There are some who believe that adverbs or adverbial phrases occurring in the middle of sentences should always be encased in commas. But these commas are not always necessary; there is a difference of meaning between the following two sentences for example:

'The bearing was apparently undamaged.'
'The bearing was, apparently, undamaged.'

The first casts some doubts on the accuracy of the inspection.

Semi-colons can be used a lot more than they are; they are useful for avoiding the common trailing-participle ending of sentences, which usually sounds so weak. Thus:

'The sodium and potassium content in the coal are the important factors, the chlorine being merely incidental.'
'The sodium and potassium content of the coal are the important factors; the chlorine is merely incidental.'

The substitution of a semi-colon for a comma and a small change in the wording has imparted more vigour to the sentence.

5.12 Report or article writing checklist

1. Choose a title—with appropriate keywords.
2. Sign the article fully.
3. Provide a summary.
4. End with an acknowledgement.
5. Total the words.
6. Cut formality to a minimum.
7. Use short words in short sentences.
8. Favour active—rather than passive—voice.
9. Eschew all but essential mathematics.
10. Eliminate all woolly words and phrases.
11. Delete all slang.
12. Follow BS560 and BS1991 for abbreviations.
13. Have the rough draft checked by a friend.
14. Have the final draft typed on one side of the paper, with double line spacing.
15. Have two copies made.
16. Number all figures in the order mentioned in the text.
17. Keep all figures separate from the text.
18. Provide useful and informative captions for all figures.
19. Supply clear line drawings, with all unnecessary detail removed.
20. Protect photographs against damage.

5.13 Further reading

A Jordan, S., *Handbook of Technical Writing Practices*, Volumes 1 and 2, John Wiley/Interscience (1971).

B Sheering, H. A., *Reports and How to Write Them*.

C Cooper, B. M., *Writing Technical Reports*, Penguin (1964).

D Rhodes, E. E., *Technical Report Writing*, McGraw-Hill (1961).

E Grieves, R. and Hodge, A., *The Reader Over Your Shoulder*, Cape, London (1967).

F Brookes, B. C., 'Report writing improves morale', *Engineering* (3 March 1961).

G Whitehouse, F., *Documentation—How to Organise and Control Information Processes in Business and Industry*, Business Books, London (1971).

H Vallins, G. H., *Good English: how to write it*, Pan (1951).

See also References 21 and 22.

5.14 Useful reference books

The acknowledged premier reference book is H. W. Fowler's *Modern English Usage* (Clarendon Press, Oxford), and every intending author should have read Sir Ernest Gower's *The Complete Plain Words* (HMSO, London). Roget's *Thesaurus of English Words and Phrases* (Longmans) and *Usage and Abusage: a guide to good English* (Penguin) are also valuable. As well as a good dictionary the Penguin *Dictionary of Science* is a useful reference to have at hand, and also the Penguin Dictionary of Quotations.

The *Writers' and Artists' Year Book* also contains some useful information concerning the preparation and submission of typescripts for articles and books.

6 Technical business correspondence

In industry and commerce there is a continual need to broadcast information giving instructions and stimulating action. Very often marketing departments deal with the specialised tasks of stimulating interest through sales letters or other advertising literature. But here we are concerned with technical business correspondence, such as letters, memos, diary notes and visit reports, etc.

Our main concern is to transmit or record information exactly or make some precise enquiry. It is essential, therefore, that a direct and specific style of writing is used.

Of course, if a piece of correspondence has to go outside the company then it will help to create a 'company image'. Everything that is written not only passes information but creates an impression of the writer and his organisation. High-flown style creates an impression of pomposity; roundabout phraseology creates an impression of tortuous thinking; impersonal, passive writing engenders a feeling of facelessness.

There is no business reason why a company's mail should be depressingly pedestrian or pedantic. Admittedly there are limitations of matter, of style, of language, but one can be human while being also concise, exact and lucid.

Talent in letter writing is a matter of caring: caring about the accuracy of what we write, caring about being of service, caring about the esteem in which your firm is held, and caring about personal reputation [26].

A business letter needs more than a good shorthand typist or secretary. She can repair broken grammatical construction and put the commas in the right places, but she cannot supply facts, or add colour, or replace muddled thought, or say what has been said in a way more likely to please and appeal to the interest of the reader.

It is imperative to write letters that interest readers, capture their attention, win their approval and get the desired reaction. When an engineer takes trouble to find out why and how his engine works the way it does, he is a better engineer than if he only knows that to pull a certain lever starts the engine. When a writer understands how his mind works, and how the minds of other people work, he will be in a far better position to progress professionally than if he knows only how to put words together grammatically.

As with all writing, consider the reader—his level of comprehension, his present understanding of the subject, any prejudices he may have, any blind spots that make it difficult for him to accept what you wish to say, and the kind of language he will consider appropriate.

Although you can't make rules for good writing because rules are substitutes for thought—and good writing requires much thinking—you can and must apply certain principles.

6.1 The Formal Aspects

6.1.1 Addresses and signatures

A good business letter will follow the general principles of any well-written and well set out letter but certain formal aspects of it are much more important than in the case of a private, personal letter.

To begin with it is important that it be written on your company's letterhead which will of course carry your address. Otherwise you will hardly be taken seriously as your company's spokesman.

It is equally essential that the address of the organisation you are writing to should appear top or bottom left of your letter, and not merely on the envelope. This should include not only the name of the company but also the person or reference—or at least function—you are addressing. Letters are taken out of envelopes in the company post room and then have to find one or two out of maybe thousands of people in an organisation.

Make sure that names and company names, titles and degrees are spelled properly: it may not seem very important but people and firms are ridiculously sensitive about the spelling of their names and titles. Unless you have a very good secretary, check also that the address is right; this can save much time.

Naturally every business letter must be dated in the top right hand corner; and preferably not a week before it is actually posted. Very important issues can hang on dates.

Unless a letter is signed in the originator's handwriting, it is of course not formally and legally valid but do not expect the chap at the other end to decypher your squiggle. Unless your writing is very clear indeed, have your name and function typed underneath the signature. The function is often very important and can give a different meaning to the entire letter.

6.1.2 The Text

If you have any say in it, see that the quality of paper and print on your letterhead is reasonably good; some people do judge your status by such things. In any case, the letter should be well set out, with plenty of margin on the left and space between paragraphs and under the headings, etc. A good secretary does this without being asked and although the recipient will probably assume that you did not type your own letter, a good secretary is a status symbol and unless you can get yours to set out your letter well, your own prestige will suffer.

This applies even more to bad spelling. There is something very off-putting about receiving a business letter full of misspelt simple words. Again you will suffer for the sins of your secretary. It will be rightly assumed that either you can't spell yourself or you haven't bothered to read the letter through.

In any case, checking a letter is quite essential unless your secretary is a real paragon. Even then you should read your letter for this, much better than your memory, will show you how it will strike the recipient. And so many a serious error of fact or judgement has been avoided at this stage. And certainly many typing and spelling mistakes.

If your letter concerns one or a few separate subjects, each should be set out under a separate heading, underlined, possibly in capitals. The reader does not necessarily know what you are talking about until he gets half way through a paragraph unless you tell him in advance. If you don't, he may miss important aspects of the earlier sentences. Or he may think you are still waffling on about rebates when you are already complaining about deliveries.

Finally, we used to be taught that 'Dear Mrs. Jones' or 'Dear Jones' were impolite unless you knew the chap personally; also for 'Yours sincerely'. Nowadays it is considered preferable to 'Dear Sir' and if you've met the chap more than once you call him 'Dear Joe' and send him 'Kind regards' or 'Best wishes'.

'Dear Sir' and 'Yours faithfully' remain for those you've never heard of.

6.2 Some basic principles of style

6.2.1 Write English and avoid Business English

Business English is a turgid mixture of heavy phrases and high sounding words. (A)

> My understanding of the position relative to the above enquiry is that . . .
>
> Your favour of 12th inst. to hand re chromium plating . . .
>
> Referring to yours of 15th Jan 1971 ref. BB/AP/194721, it has been assumed for the purpose of this reply . . .
>
> It is recommended that efforts should be made to obviate this difficulty by . . .
>
> It would be appreciated if you would favour us with your comments . . .
>
> Awaiting the favour of your further advice. . . . .
>
> I believe that this point requires the earliest possible clarification as it reflects back on the methods of obtaining costs. . . .
>
> It has been assumed for the purpose of this preliminary investigation (into availability) that spares are readily available since this assumption avoids the indeterminacy associated with replacement items. . . .
>
> Further to the above, with particular reference to your payment of the 2nd Feb. last when you sent us a cheque for **$160.02,** we would explain that in the interval which has meantime elapsed the papers herein have unfortunately got mislaid in the process of filing. In the near future and in the ordinary course, you will, doubtless, be favouring us with your further remittance in respect of . . .

and so on, and on, and on.

Remember also that not all unnecessary words are long ones. Why, in our business correspondence, do we still all too often beg to thank you for your letter of the umpteenth ult. or inst.? Are we really begging? And, incidentally, why 'ult', or 'inst'? What is the matter with a good old-fashioned date? Why do we assure people of our best attention? Do we have several kinds of attention, the best being reserved for certain customers? One thing is certain: if a word is in any way misleading, or if it means nothing at all, it is better left out [27, A, C].

6.2.2 *Know what you want to say*

A business letter must either enable a transaction to be effected without personal contact or provide a record of all the significant factors of a transaction, or merely give data and information. The writer must therefore know what has to be said and the sequence in which it is going to be put. Write the way a structural engineer builds—first he drafts his plan and then designs every detail from the foundation upwards.

6.2.3 *Express it briefly*

Confine your words to those in common use and avoid the ponderous business English mentioned above. Remember what George Eliot said: 'Blessed is the man who having nothing to say abstains from giving in words evidence of the fact'.

Here are some words and expressions which must be abolished from technical correspondence.

Phrases to avoid	*Say instead*
Your esteemed favour	Your letter
Yours to hand	Thank you for your letter
Advertising to your favour of	Referring to your letter of
Enclosed please find	I (we) enclose
We have received same	We have received it
Please be good enough to advise us	Please tell us
Awaiting the favour of your esteemed command	*Omit entirely*
Assuring you of our best attention at all times	*Omit entirely*
We regret to inform you that we are in error	We are sorry for our mistake
Trusting this matter will have the attention of your good selves	*Omit entirely*
Only too pleased to	Very glad to
At an early date	Soon
Awaiting a favourable reply	*Omit entirely*
We beg to remain	*Omit entirely*

Ult; inst; prox;	*Give the month its name*
Without (as a conjunction)	Unless
Same (as a pronoun)	It

See Appendix 6 for some further examples and Appendix 12 for the misuse of words. Also use a short word wherever possible to express your meaning:

Long	*Short*	*Long*	*Short*
acquaint	tell	proceed	go
assist	help	purchase	buy
commence	begin	request	ask
consider	think	state	say
despatch	send	terminate	end
inform	tell	utilise	use
peruse	read		

6.2.4 Be careful when using adjectives and adverbs

Don't use adjectives and adverbs unless they will increase the reader's understanding—consider this:

> 'We are hoping to despatch a relatively substantial percentage of your plutonium at a comparatively early date.'

This tells the receiver absolutely nothing, except possibly that his firm are not likely to fulfil your requirements.

6.2.5 Use the short preposition

'Our method of stress calculation is devised for speed'
not
'In connection with the subject of stressing our practice with regard to calculations is devised with a view to speed.'

6.2.6 Avoid the use of abstract nouns where possible

Not this:

> 'The conditions of heat treatment have a bearing upon the degree of deformation possible'.

but this

> 'The deformation possible without fracture varies with heat treatment given.'

Mr Hammond's Diary Note

Tuesday, 29 December 1969
E & R Office
Present Mr Lockett ABC
Mr Hammond E & R
Mr Simms E & R
Mr George E & R

Subject Radar aerial

Mr George explained that the design of the cantilever had been approved by Stress Department and the layout made by E & R was handed to Mr Lockett to be redrawn to incorporate the aerial feeder. The shape of the doubler plate at the end of the cantilever (see sketch) was agreed by all present.

It was agreed that the finished drawings would be marked-up GA's containing enough information for E & R to make the job without requiring details, in the same way as the experimental model had been handled.

The connecting rods were agreed to be made from hiduminium castings; Mr Lockett's design will be handed to Mr Hammond for stressing.

Copies to: Messrs Whitaker, Barnes, Fox and Lockett

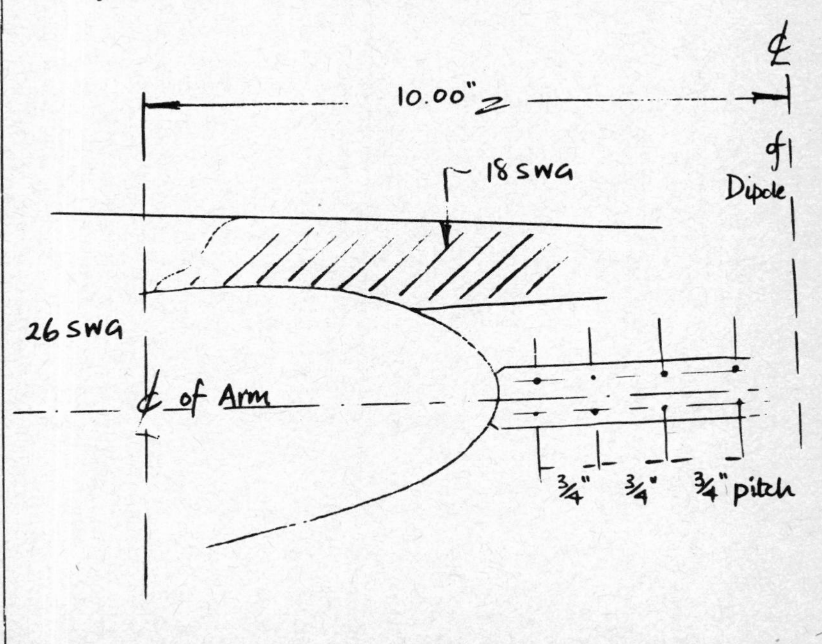

Figure 14 An example of a diary note

Whether they realise it or not many engineers and scientists shroud themselves in mystery. This is partly to make their work appear more important and complex than it really is and thus

Writing simply	is equated with	being simple minded
Mathematics	is equated with	quality and value for money
Pompous language	is equated with	dignity
Complexity	is equated with	wisdom
Formality	is equated with	importance

Remember it is not a sign of weakness or defeat if a typescript ends up subject to major surgery. This is a common occurrence in all writing, and among the best writers (see Fig. 11 and Section 5.7).

6.3 Diary notes

In engineering, especially in R & D and design work it is sometimes necessary to record decisions taken at informal meetings, say in a laboratory or on a drawing board. A diary note for the file is a useful way of doing this. Its format should be simple and give the date, time and place and the people present when the decisions were taken. The heading of the subject should be chosen with care. This should be followed by a simple but clear statement of the decision reached and if necessary, sketches or prints of drawings attached.

A typical example is given in Fig. 14. The main aspect to note here is that the points of agreement are clearly stated and the actions agreed are recorded.

6.4 Research and design data

Although a special case in business correspondence, printed information for engineering designers is of vital importance. Often this is poorly written and presented. It is essential that in the production and preparation of information for engineers the following should be considered [28, 29]:

1 Easy storage, retrieval and handling.
2 Good indexing and cross-referencing.

3 Logical order and consistent format.
4 Clear and pleasing presentation.
5 Appropriate degree of detail for intended readers.
6 Precision and accuracy of quantitative data.
7 Inclusion of publication date and provision for updating.
8 Indication of sources of further information.
9 Suitability for use abroad.
10 Effectiveness with economy.

Out of 220 submissions to the Design Council's panel for display at their 'More Value by Design' exhibition, only 40 were chosen and no submission satisfied all ten criteria given above.

6.4.1 Nomographs

In presenting technical data, consideration should be given to using nomographs as these are time-saving devices for users [29]. Some care, however, may be necessary in constructing the nomograph to make it easily readable. A good example is Fig. 15 which can be used for finding the second moment of area of a beam section about any plane parallel to the neutral axis of the beam:

$$I_{xx} = I_{CG} + AK^2$$

where I_{xx} is the second moment of the area required, I_{CG} is the second moment of area about the neutral axis, A the cross-sectional area of the beam and K the distance between planes CG and xx.

If $K = 2.0$ in and the area is 4.6 in^2 and I_{CG} is 32 in^4 then by drawing the line AB (4.6 on the A axis to 2.0 on the K axis) and then drawing line CD from 3.2 in^4 on the I_{CG} axis at 90° to AB the second moment of the area about xx is obtained.

6.5 Special types of business report writing

In addition to general business correspondence in the form of letters and memos, etc., there are a number of other special technical reports that have to be written from time to time. While the basic principles outlined in Chapter 5 still apply there are some general points which should always be kept in mind.

6.5.1 The visit report

Technologists are constantly visiting firms, construction sites, manu-

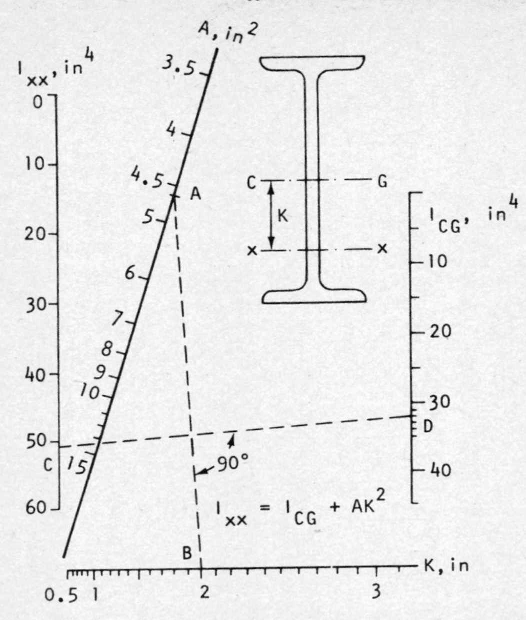

Figure 15 Nomograph for finding the second moment of inertia of a beam

facturing concerns and exhibitions. It is important that good records of such visits are maintained. Too often such reports are written in the time sequence in which they occurred; it is, therefore, difficult on reading them to infer the main ideas and even the purpose of the visit.

Put the 'what', 'where', 'when', 'who' and 'so what' first, then follow this up by the details expressing the 'why' and 'how'.

The structure of such reports should be rather like a newspaper report (see Fig. 16). Here the 'where', 'who', 'what', 'when' is followed by the 'why' and 'how' with the necessary build up in details as required. This type of structure is necessary for a newspaper article since the editor may need to 'cut' because of shortage of space. He can do this without having to rewrite the article all over again. This technique can also be useful if an engineer wants to send details of a visit report to top management. He can cut at an appropriate point for the managers and send a full copy giving details of the 'why' and 'how' to his peers. The structure is different for a magazine article. Here a lead into the subject is required with a build up to indicate significance which is followed by development and projection as outlined in Fig. 17.

6.5.2 The monthly progress report

Of all writing jobs in technical laboratories and design offices the monthly progress report causes more grumbling than almost anything else. Is such a report necessary? This must be answered by the manager,

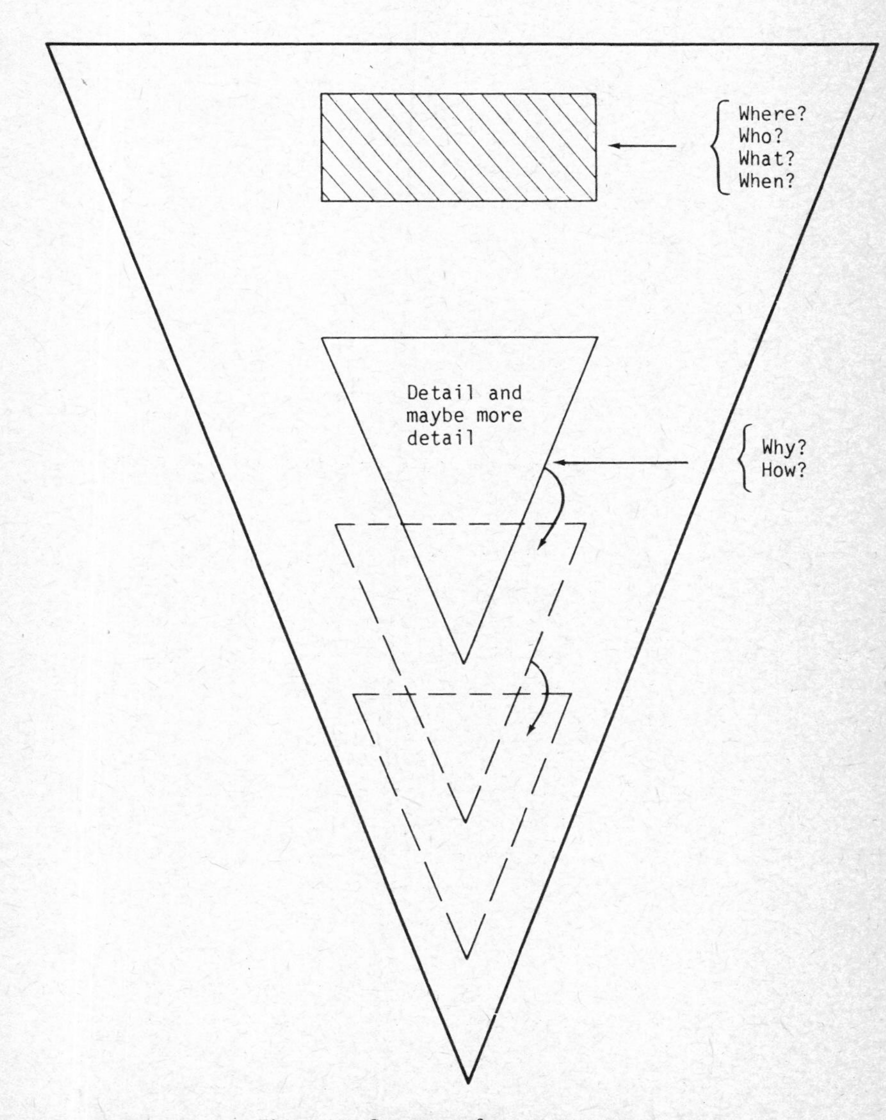

Figure 16 Structure of a newspaper report

who should explain why he needs such a report. Is it to inform, guide or just record progress?

It is essential that such reports state problems fairly and do not strain to give results that are not yet achieved. A good deal of time can be saved if the format for 'on-going reports' is clearly set out with wide

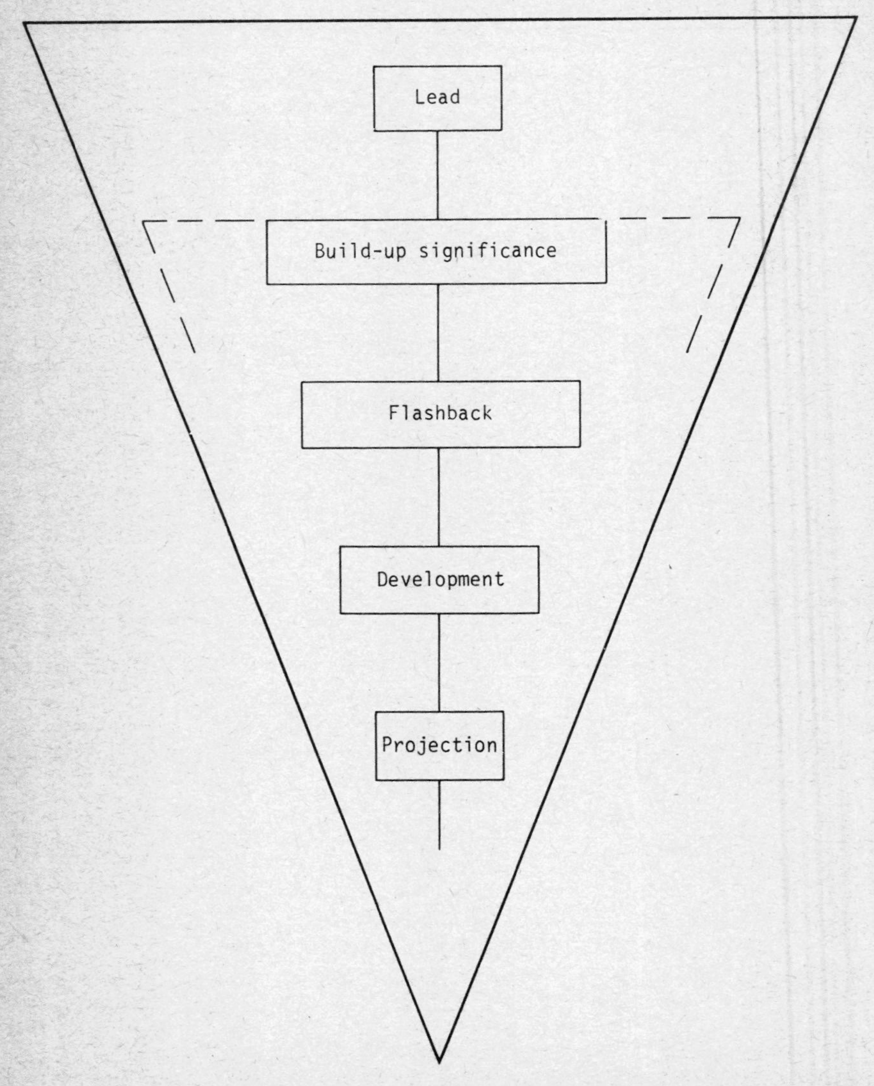

Figure 17 Structure of a magazine article

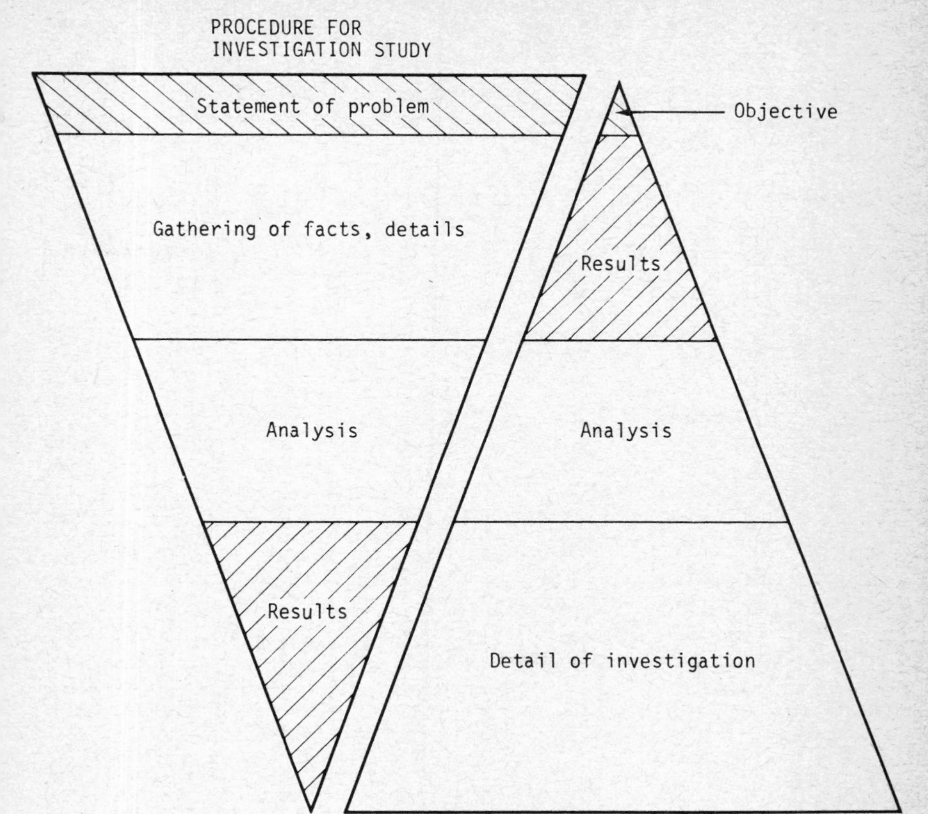

Figure 18 Comparison between an investigation study and a report

margins to list names of people who should take action on the various subjects. A good clear diagram (bar chart or arrow network) should be used to save writing reams of dates, etc, too.

6.5.3 The investigation report

Management frequently assigns an engineer to investigate a particular problem. When the engineer reports back very often he does so in the same way that he has carried out the study. The structure of his report should help the reader and save him time. The investigation will have started with a collection of facts and then to analysis and so to the final results. But management will be more interested in results first and so the presentation of the written information should be structured as indicated in Fig. 18.

6.5.4 *Procedure manuals*

Large engineering organisations find today that with many products being produced there is a need to regularise and formalise procedures. This can be a tedious technical writing job but when completed forms a useful 'bible' for management. Procedure manuals are documents that concern action rather than ideas. A typical example would be the procedure for design or manufacturing processes. Under this heading are included such items as modification procedures, approved component control, design aims, dimensional control, signing of drawings, etc.

Here the writer must be very familiar with the firm's existing operations and be able to write down clearly each step of a procedure. A typical example is given in Appendix 13 of the required format and contents of a test specification. In this particular case it will be seen that the document can act as a useful checking list for the designer concerned.

6.6 Further reading

A Kirkman, A. J., *Making words work for you*, paper presented at PERA Symposium (November 1966).

B Lancaster, J. B., *The economic value of technical information*, paper presented at a Conference held by the Institute of Information Scientists (May 1969).

C Shurter, R. L., *Business and Research and Report Writing*, McGraw-Hill (1965).

D Sidney, E., *Business Report Writing*, Business Publications (1965).

E Machon, F., 'Information for engineers', *Engineering Materials and Design* (January 1961).

F Bell, R. W., *Write What You Mean*, George Allen & Unwin (1900).

G Price, S. S., *Business Ideas*, Harper and Row, New York (1962).

H Carey, R. J. P., *Finding and Using Technical Information*, Edward Arnold, London (1966).

See also References 28 and 29.

7 Technical talks and how to give them

Very few men, or women, are able to stand up and speak to an audience, whether to propose a toast, to lecture to a class or to deliver a paper, without initial discomfort and nervousness. Let us all take comfort from a saying by the poet Goethe that 'fools are never uneasy'. However inhibiting attitudes and tensions can do much to stalemate our progress in speaking. Inhibition wears many masks. It may be manifested in a flat monotonous inexpressive voice, or an expressionless face or a rigid posture. The most sensitive speakers are too conscious of their faults and limitations. They live in the negative and fail to move over to the positive and remember their good points. After all technical people are on the whole better informed about the material world than most of the population. They need not be afraid of having nothing of interest to say. Nervousness can only be overcome by creating a positive attitude of mind and practising and practising again. Failure and disappointment are an inevitable part of the learning process. With practice you will begin to succeed and to boost your self-confidence so that when disappointment comes your way you will have the strength to persist. Then and only then will you break through the barrier of persistent inhibitions and get into the upward spiralling cycle of self-confidence, less inhibition, more successful speaking.

Most of the remarks made previously about written communication apply to technical speaking. There are however important differences such as immediate feedback as the speaker sees his audience and they ask questions. If there is a captive audience the length of the message may be crucial! The whole process of speaking makes communication more personal and the sender tends to convey unspoken thoughts by his actions and mannerisms as well as the tone of his voice. These added dimensions of presentation must be carefully considered.

The speaker who has fifteen minutes allocated to him must necessarily prepare differently from the speaker who has been allocated an hour. Curiously enough if the subject is to be treated properly the shorter opportunity demands the longer preparation. This also applies to writing for as one editor said 'It will have to be a long editorial tonight as I've no time to write a short one!'

7.1 For the informal meeting

At an informal technical meeting the author is able to use much greater freedom of speech. Use of the first person makes expression much easier. If, after speaking to a learned society, the paper is published it will generally be transcribed into the third person, and sometimes the author will be referred to as 'the speaker'. Once committed to print, a statement has to stand the hard searchlight of criticism at a distance. With the spoken word there is the possibility of modification by intonation and expression followed by immediate explanation, or by expansion of the idea in the discussion. For the audience, the easier form of speech, using the first person is less tiresome. An informal speech must not last more than forty minutes. Much better to be short and crisp than long and boring, leaving sufficient time for discussion. The time available for discussion also gives the speaker an opportunity for correcting misunderstandings and for enlarging on points which he previously considered were obvious.

For an informal meeting the actual words need not be written in detail, though many have not the skill of speaking without preparing full notes in advance. Sir Frank Smith once spoke at considerable length with great fluency and no notes; afterwards, someone observed how much those present had been impressed by his easy delivery. His reply was: 'I should consider it very rude to appear before an audience without knowing exactly what I proposed to say and how I meant to say it'. That, of course, is the attitude of the experienced perfectionist, not expected of someone making his first informal speech.

Remember too Demosthenes did not consider it beneath his dignity to allege want of preparation as a reason for refusal to speak, and many an orator since his time would have best served the interests of the public and of his own reputation if he had followed this example.

Confidence and skill will grow as one speaks. Skilled orators, in most cases, prepare in advance, impromptu remarks included. It does not necessarily follow that one reads all that has been prepared. Generally it is best not to read at all but merely refer to notes from time to time. Brief heading notes can be valuable. A useful tip here is to put each heading on a separate postcard and turn them over as you speak, but be careful how you handle your notes. Indeed, if the paper is to be published it helps to prepare two copies, one for presentation to the journal for consideration and subsequent publication and the other for use at the meeting. The two versions need not agree at all in construction and, anyhow, once one is installed at the podium, it is possible to get away with quite a different version. Most speakers of technical subjects will be able to enhance their presentation by using illustrations—slides or short films can help to add impact. More details of these aids are given in Chapter 9.

7.2 For the formal meeting

For formal meetings the use of the third person is generally best but avoid reading the paper. Reading aloud a report or paper intended for printing is never effective communication. All words must have their precise engineering meanings and the same word or symbol must mean the same thing throughout the delivery, otherwise the audience loses the thread of the argument while unravelling the ambiguities.

For both types of talk it is essential to be very clear about its purpose, for the target you are aiming at will affect both the material in your talk and the way you present it. Also as already mentioned in the opening paragraphs of this manual in preparing for either informal or formal talks, every speaker must bear his audience in mind. What is their general attitude likely to be? Is it going to be one of interest, scepticism or indifference? How much will they already know? If the talk you are to give is part of a series, what have the other speakers said? How fast can the listener receive the information you give him?

7.3 Putting the talk together

Successful talks have good openings. Such openings are attention getters and audience orientators. The good speaker shows what he knows with his very first sentences. He knows what he is talking about because he has first organised his thinking. He is in control of an idea. He has investigated and explored his concept from every angle. The best speakers are always in control; they know where they are and where they are going. To do this requires planning.

The shape of your talk must be known. A well thought out plan and outline will be the shortest route from the purpose to the goal.

Your introduction must capture the interest of the audience and you must lead them on into the development of your main theme. Each step should be made memorable. Good talks often register because apt use is made of suitable illustrations, analogies and experiences. A speaker should be able to explain his main idea in a few brief sentences. The progression of his talk must give direction and maintain continuity of thought. All contributing ideas are tributaries feeding into the main stream. They only serve a useful purpose if they enable both speakers and audience to travel efficiently in the right direction to the objective of the talk. The rhythm of a talk is also important. It must be balanced, not too crowded with too many ideas.
The pace should be continually changing to capture interest and allow adequate time for digestion of what is said.

Finally, if a good opening is essential to a successful talk a summary recapping in cogent terms of the main points of the presentation is equally important. A restatement of the main points, briefly, is always worthwhile but if a speaker requires action a vivid illustration or dramatic quotation may be useful. Never add new information at this stage in your talk. Confine your closing remarks to simple terms of your major premise and any proposed action.

7.4 Estimation of time

A word should be said about time. So far as the text of the paper is concerned a speed of about 110–120 words per minute is a rough guide. In addition, time must be allowed for slides, or other illustrations if used. If any detail at all is to be picked up, a slide must be given from half a minute to a minute and if a description is included while a slide is on exhibit, it will take much longer—say half a minute plus 100 words a minute. This time must, of course, be included in estimating the time allowed for the paper.

7.5 Value of good delivery

Consider the question of delivery. How often is a good and interesting paper spoiled by bad delivery! It may appear afterwards in a journal, but that is no justification for spoiling the evening's enertainment and it is exceedingly trying for the audience if the effort to catch the words leaves no energy for appreciating the meaning.

The ideal is a good speaking voice, which includes articulation, enunciation, correct pronunciation, pause, emphasis, punctuation, accent, rhythm, intonation, expression and feeling. One of the most common defects is reading the whole paper. Those who are lazy invariably read. Reading tends to produce a tone that is wearisome for the audience and if you cannot memorise the best thing is glance-reading, which can approximate to informal speech. Everyone tends to speed up and raise his voice while reading aloud. Both are a mistake.

Clear enunciation in not a matter of any particular accent whether of Oxford or Edinburgh, so long as the address is clear, intelligible and orderly. Slurred speech is as trying as smothered speech.

There are tricks in getting the voice away. The bull-voiced sergeant major puts his voice over by sheer strength. An officer with less lung power can put his orders through by the adoption of a nasal tone which, although not beautiful, can often be effective. The Americans are supposed to be good public speakers, they mouth their words more than the English, and in photographs always seem to be advertising the success of their dentists! It is probable that nasal resonance has much to do with their success. It ensures that their words are heard. A small voice carries quite well on a nasal accent. (For further details see Appendix 7.)

The acoustics of a building sometimes have to be studied. It is said that when preaching in St Albans Abbey it is necessary to speak towards a nearby column. So, and only so, will the sermon be heard by the whole congregation. Was it designed thus? Did the architect have in mind the dullness of so many sermons? Most of our historic churches have some such feature. Meetings in small rooms present little difficulty but it is fatal to mumble to the front row. Always look up and address the back row. There is another good reason for doing this. Addressing the people in the back row enables the speaker to watch their reactions. If they appear to be nodding, it is possible to cut out a bit or deliberately make a mistake, according to circumstances. While, if they all appear interested, it is certain that the front row is not missing much wisdom.

7.6 Avoidance of mannerisms

Linked with good delivery is avoidance of mannerisms. Clear delivery is vital to success, while most mannerisms are merely annoying. Everyone knows the sort of thing—stroking the bald patch at the back of the head, putting on and removing spectacles, a hacking cough at every third word, and so on. These are, generally speaking, nervous complaints. If they can be restrained it is well worth the effort, unless the effort is so great as to break down the delivery altogether. These remarks are not meant to offend any individual and it is hoped that nothing written here will restrict future speakers. Everyone has to overcome such defects.

In the works of that remarkable man, John Wesley, are some directions on the subject of preaching. He had to train men and he dealt with enunciation and voice production. He laid emphasis on modulating the voice to suit the room, speaking neither too loudly nor too softly and, in particular, avoiding the common tendency to drop the voice at the end of a sentence, a fault which deprives the audience of the most important words. He also spoke of some mannerisms which have happily nowadays been discarded. It is interesting to quote his advice. 'Labour to avoid the odious habit of spitting when you are preaching and if, at times, you cannot avoid it, yet take care you do not stop in the middle of a sentence, but only at such times as will least interrupt the sense of what you are saying.'

7.7 Allow spare material

It is as well to have a small reserve, or to leave some points unsaid. If the discussion flags, these can be worked in by use of a little guile while answering questions. If these paragraphs are necessary, they can always be included in the published paper, and indeed, it is well to go through the paper as prepared and mark some of the paragraphs for omission if time is short. Failure to give them will probably raise a question and thus promote the desired discussion. It is like the device of leaving a small mistake if you want to get a scheme past a critical senior; he is so pleased at finding the small error, that he swallows the major portions of the scheme [30].

7.8 Conference and convention talks

Today it is quite common for engineers and scientists to be asked to speak at conferences which may well take place on an international scale. Indeed they have become the modern market place for exchanging ideas. Invariably the programmes are too crammed with papers and the attendance too large.

It is as well to be prepared to have your speaking time severely curtailed unless you are the first speaker. Most audiences get extremely weary after sitting for an hour or so listening to the opening address and preliminary announcements. If you have to follow in this situation make sure your are brief and only give two or three main points from your paper.

There are two additional problems: one is following a brilliant speaker and the other is following a poor speaker. Unfortunately there are all too few of the former but if you do have to speak after one, again be simple and brief and use, if possible, some good visual aids. Your audience will be well tuned in by the time you are on and it is important that you hold them by maintained interest but be your natural self and adopt your normal style. Never try to copy the previous speaker.

The poor speaker is all too familiar. Quite often he drones on and on reading large chunks of their published paper which has already been issued to delegates. By the time it is through—and he generally exceeds his time limit in spite of the Chairman's interjections—everyone has had enough. At this stage you are called upon to speak. As you get up people may be leaving the room—it's all very depressing. Unless you are very good, don't start with a joke but tell them straightaway that you have taken your paper 'as read' and you intend very briefly to highlight three main points. These should, where appropriate, be augmented by visual aids. Having made your points, repeat them and mention advantages if any and sit down. Remember most technologists try to give too much detail and too many facts to their audiences. If it can be done in ten minutes you will nearly always find you will get applause.

I once had this experience at an international conference and gave my talk in seven minutes flat. During the tea break many delegates congratulated me on a smart presentation, but the next day the complete text appeared in the daily paper. When I enquired why my paper was printed and no one else's the reply was: 'You were given no time to read the paper so we felt it should be published in full.'

One final point, remember that if delegates come from various countries and simultaneous translation is used you need to watch idiomatic phrases, colloquialisms and abbreviations. With good translators there is no need to speak more slowly unless you are by nature a fast speaker. Sequential translation is a very different situation and of course, cuts down the amount of material you can present, but gives the speaker ample opportunity to observe audience reactions. It also allows the speaker time to arrange his thoughts and adjust his slides, etc. If it is possible to discuss your presentation with the translators before the conference this is well worth doing.

7.9 Seminar talks

A seminar is a very different kind of meeting to the conference or convention. Here the leader will be putting forward new ideas or new slants on old ideas. He may, and should, put questions to his audience and must expect interruptions. The meeting can be much more informal but the speaker has to act as chairman as well as giving the talk. In other words he must keep to the programme laid down and therefore will need to clamp down on side issues and any 'red herrings'. He should allow and sometimes encourage cross-talk between members of the audience. One of the arts of running good seminars is designing suitable exercises that can be done in syndicate groups so that the main points of the speaker, or speakers, are driven home. The style of speaking to be adopted here is somewhat different from that at a convention with its platform. Indeed, it is often possible for the speaker to move into the audience to emphasise points and ask questions. Nevertheless there is still the same need for clarity of expression and good articulation. It is the element of participation that differentiates a seminar from a formal meeting. Skill at handling questions is the most important aspect for the speaker at seminars.

7.10 A rapporteur's talk

Sometimes at conferences where there are a number of technical papers to be given the organisers appoint rapporteurs to cover several papers at once. This is a very effective way of getting the best presentation of a wide coverage but is very hard work for the chosen rapporteur. He must read and thoroughly digest all the papers and avoid injecting his own views. To do this well he must have the ability to distil the main ideas and join them together setting out clearly their relevance and importance. It goes without saying that to do this he must be well versed in his subject and have a good analytical mind. He must also try to maintain impartiality but yet evoke a worthwhile discussion. Before taking on such an appointment any speaker should make sure he has had ample time to prepare and for this reason the papers must be available well before the conference is held.

7.11 Speaking on television

In the future it is possible that more and more engineers and scientists will be called upon to speak on television about their work.

There are important differences when it comes to speaking in front of a television camera. The greatest difference is that the speaker has no visible audience. He can't therefore gauge the response and this lack of feedback can be disarming. To be successful a speaker needs to adopt a technique similar to the film actor who knows exactly how he is going to appear on the film. One way of doing this is to use a closed-circuit television (CCTV) system. Here the would-be speaker can see himself as others see him: he can become both speaker and audience. A glance at the monitor screen whilst speaking will quickly reveal any errors.

Another important difference between television and other forms of public speaking is one of character. The screen focuses the whole attention of the audience on the speaker. Every mannerism and movement is exaggerated. This is true of both the picture and the voice. Mannerisms, whether of gesture or voice, become irritating and distract attention from what is being said. Undue movement too, is distracting. The microphone picks up a very wide range of sound and draws attention to faults in speech and fidgeting until in some cases the result is laughable.

Because television presentations show a speaker close up and tend to make him appear pale it is often necessary to have make-up applied. The human face when seen by a television camera without correct make-up will lack definition of such features as nose and mouth. The shape of the face will appear uninteresting and flat without proper moulding to cheekbones and jaw lines etc. All these points need correcting by suitably applied make-up. This is an expert job and speakers should always accept the offer of having make-up applied by an expert if they have to speak for any length of time.

Apart from these points, television speaking does not differ from ordinary public speaking at a technical meeting. The main points to remember when speaking on television are:

1 Know your subject.
2 Rehearse before delivery.
3 Face the camera being used (indicated by light or otherwise).
4 Focus your eyes on an imaginary person at the camera.
5 Speak slowly, clearly and carefully (125 words per minute maximum) using a lively conversational style.
6 Sit still and avoid sudden movement.
7 Avoid mannerisms of speech or gesture.
8 Do not gesticulate unless it is natural for you to do so.
9 Be precise in your timing.
10 Be relaxed and at ease.
11 Look at the camera and not at your notes.

Remember the great skill is to be still, yet not wooden, relaxed but not slumped.

7.12 Preparation

Before reading a paper or delivering an informal address, it is as well to try it on the dog or in other words, to inflict it on a candid but well disposed friend. One is under no obligation to accept advice if it is not liked. Many men try out their speeches on their wives who can usually be relied upon for candid criticism and patience [30]. For technical people it is often possible to have a dummy run with your own staff at the office. A small group who have been concerned with the work in one way or another can be very helpful in giving a speaker an opportunity to test his message and method of presentation. But such efforts are only experimental and different from an actual performance [30, 31, 32].

7.13 Security restrictions

Restrictions may be imposed by firms who are unwilling to have details of the work made public, either for commercial reasons or for the sake of their reputations, though as a rule firms are very reasonable and some may see in the preparation of a paper evidence of the keenness of their staff or more possibly good advertisement. In the U.K., the Official Secrets Act may be enforced. Such restrictions must be respected and permission obtained before the paper is read. Any premature disclosure, whether by the main speaker or in discussion should be notified to the appropriate person immediately, who will contact the editor so that any undesirable matter, whether in the paper or the discussions, is not published.

7.14 Message completed

Speaking is a performing art. All speakers have to learn how to perform with economy of effort, poise and purpose. You have to practise and go on practising. Don't apologise or denigrate yourself. If you have a proclivity for humility no one is likely to defend you. They are much more likely to switch-off. On the other hand too much self-confidence, self-assurance and an easy nonchalant manner can also be damaging, especially when self-confidence turns to patronising superiority and assertiveness to dogmatism. Learn to steer between the Scylla of humility and the Charybdis of over-confidence.

7.15 Checklist for preparing and giving a talk

7.15.1 Preparation

What is the objective?
- Teach
- Persuade
- Stimulate
- Entertain

What do I know about the audience?
- How many
- Background, experience, knowledge
- Age, status, sex
- Knowledge of subject
- Reaction to subject, and to me.

What do I do about the situation?
- Size of room, seating, etc.
- Lighting, acoustics
- Time available
- Facilities—visual and audio

About four weeks before the talk, spend time writing down some of the topics that might be appropriate. Perhaps discuss it with other people.

Then let your sub-conscious mind handle it for the next fortnight or so, and you will be surprised how much work gets done for you.

Collect material from all sources:
Brainstorming
General reading
Talking to other people

Detailed preparations:
Select the appropriate material
Identify the main points
Plan the supporting narrative and examples
Arrange in logical sequence
Prepare 'aids and props' which will help
Plan the opening and closing
Time the talk
Make brief notes
Practise the talk

In this way you will approach the audience in a state of preparedness which will give you confidence, and which they, as your listeners, deserve.

7.15.2 Presentation

1 Establish a relationship—introduction
Gain their co-operation—participation
Interest them—impact of opening
Relax, and be sincere
2 Choice of word: Descriptive
Arresting
Appropriate and accurate
Emotional
3 Sentences: Variety
Colourful
4 Use of voice: Pitch
Pressure
Pace
Pause
Variety
5 Movements: Look at audience—all of them
Calculated and controlled

Exaggerate demonstrations
Mannerisms
6 Participation: Questions
Exercises
7 Summaries and conclusions
8 End—and finish. Don't apologise.

7.15.3 Some further points

Most amateurs:
Mumble
Talk too fast
Bring too much material
Never stop to invite comment
Never ask questions
Can't finish
Don't use aids

Aids to confidence:
Preparation and practice (rehearsals)
Props and visual aids.
Forget yourself
Try to like and enjoy your listeners (or appear to!)
Be sincere.

7.16 Further reading

A Carnegie, D., *Public Speaking*, Cedar Book (1957).
B BACIE, *Tips on Talking* (1961).
C Whitting, B. H., *How to Speak and Write with Humour*, McGraw-Hill (1959).
D Howard, V., *Talking to an Audience*, Sterling, New York (1963).
E Gandin, W. R. and Mammen, E. W., *The Art of Speaking Made Simple*, W. H. Allen, London (1967).
F Elliott, S., *Quick Guides to Speaking in Public*, Mercury House (1972).

See also Reference 30.

8 Running meetings

The British are said to have a genius for committees and, indeed, much of the official and voluntary work is undertaken by such methods. Achievement, if there is any, depends upon discussion and consensus as opposed to dictatorial methods.

There are many definitions of committees not all of them very kind, as, for example:

> 'A committee is a noun signifying many but not much.'
>
> 'A committee is a group of the unwilling called together by the ignorant to do the unnecessary.'
>
> 'A committee passes minutes but wastes hours.'

Professor MacKenzie defined a committee as: 'A body of people meeting round a table, to take decisions for joint action on behalf of some other (generally larger) body of which it is the committee' [33].

Presumably the process of decision-making is by discussion and that each committee has some terms of reference. Committees can be classified in various ways. There are, for example, official institutions' standing committees which meet at regular intervals to discuss various problems. There are also *ad hoc* committees or working parties set up to study particular problems. Having solved the problem they are disbanded. Then there are executive committees, these have power to make final decisions and enforce them, whereas advisory committees, as the name implies, merely have to advise and provide a platform for airing views. Generally in engineering the technologist and engineer is more concerned with Staff meetings, although from time to time particular engineers may be called to serve on special government committees, or committees of a professional institution. Staff meetings form important communication channels for industry and the technical profession.

A meeting may be said to be an assembly of three or more people whose objective is to achieve, by discussion and decision, something which cannot be achieved by the people acting in pairs, or as individuals. It may be structured in a very formal way, or be a casual meeting of interested people, convened on the spur of the moment, or anything in between. Whatever type of meeting it is, it is likely to achieve better results if it is guided skilfully towards its objective by the chairman, or leader; and the members of it are co-operating objectively.

Everyone taking part in a meeting must realise that the outcome depends upon the spirit in which they handle the problems before them as well as the skill and force with which they put their own point of view.

Meetings will more easily find the best solution when all the knowledge available is pooled. Personal pride, prejudice, private squabbles, private distractions, shyness or self-aggrandisement will hinder the meeting. In addition to mutual sharing and interchange of information, meetings should serve to clarify, stimulate and sharpen participants' thinking.

8.1 Types of meetings

There are various types of meetings as, for example, the following:

1 *Learning situation*—Here a manager or team leader may be trying to obtain feelings, attitudes, ideas from the group or he may actually lecture to the group and invite questions or comments.
2 *Getting a group decision, or making a plan of action*—In this kind of meeting participation is the keynote, the group puts forward suggestions which are discussed and a possible plan of action drawn up. Differences are resolved and a consensus obtained.
3 *Problem solving*—A typical type of engineering meeting is one concerned with solving a problem. Various techniques such as brainstorming, synectics, or value analysis may be used for working out possible solutions (C, D).
4 An increasingly important type of meeting is that between engineers representing management and trade union representatives. Today success or failure in industry often hinges on industrial relations which differ from all other types of meetings in being neither technical in the strict sense, nor non-technical. They often concern technical aspects of shop-floor work and make very great demands on speakers who have to explain technical points that are relevant.

It is important for any technologist to be clear about which type of meeting he is attending or convening.

8.2 Some initial factors that must be settled

What is the purpose of the meeting?
What are the objectives?
Who needs to be there?
Would interviews be better?
When is the best time to hold it?
What factual information must be available before, or at the meeting?
Remember in industry and certainly in government establishments—people like meetings. They are cosy: the speakers have a captive audience. It is not hard work (except for the Chairman).
The political intrigue is exciting.
They can be very time-consuming [34].

8.3 Some points to remember

1 *Every meeting has two agendas*: (a) the formal—known, formal working agenda or objective and (b) the informal—hidden objectives motivated by interpersonal or interdepartmental struggles and misunderstandings. (Remember the agenda is the route map for any meeting. It determines the subjects to be discussed and their order of consideration.)
2 *Insecure environment*—In an insecure environment, decisions are arrived at on the basis of avoiding open conflict. Sometimes people don't know what they are deciding upon, or voting on. Often executive action is not specified.
3 *Inhibition*—Juniors may be inhibited by seniors. Contributions may therefore be unbalanced.
4 *Repetition*—Time can be wasted by covering the same ground again and again.
5 *Timing*—The group can lose interest and stop working, especially if the meeting starts late or goes on too long.

8.4 Behaviour at the meeting

1 *The chairman or leader—*

a defines the limits of the discussion;

b is impartial, but gives some pros and cons to get people thinking;

c controls the discussion and speakers;

d makes frequent interim summaries;

e avoids arguing: throws controversies back to the meeting;

f draws conclusions, sums up at the end and states action to be taken;

g thanks the members.

2 *The members—*

a keep within the limits set by the chairman;

b discuss everything objectively;

c try to find solutions to problems;

d be simple, brief and clear.

8.5 Control of the meeting

1 *Before the meeting—*

a invitations must be given to the right people and the objective clearly stated;

b an agenda should be prepared and where possible sent to all who are to attend;

c notices, handouts, drawings, factual information must be checked and collated;

d members should be briefed where necessary;

e arrangements for minute taking should be made;

f physical comfort should be arranged—not too comfortable but good ventilation essential.

2 *At the meeting—*

a start and finish on time; welcome and give thanks;

b keep stating objectives clearly, and summarising results;

c use the statements and questions that you have in your plan (see Section 8.8);

d vary the order in which you ask for contributions;

e keep the tempo going;

f if there is disagreement, sum up the points of agreement;

g control lengthy speakers and encourage shy ones;

h if there is to be action, state clearly who is responsible for its execution;

i minutes—notes should be taken as the meeting proceeds—a small tape recorder is best for this purpose.

8.6 Plan of the meeting

The chairman will have analysed the subject, seen the problem and divided the problem into manageable sub-sections, to be solved one at a time. He will have outlined the way the meeting might go, but this plan must be flexible in the light of contributions from members.

To guide the meeting, and to provoke thinking, he will prepare:

1 General statements.
2 Specific statements.
3 General questions.
4 Specific questions.

He will have considered his members' reactions and planned the approach—where to encourage and give credit—where the knowledge may lie—whom to control etc.

8.7 The minutes

The minutes are the only record that the meeting has taken place and decisions have been reached. They can be a launching pad or a diving board. They must therefore be:

1 Factual, impartial and balanced.
2 Present a clear, concise and unambiguous record.
3 Brief.
4 Well set out to aid assimilation of contents.

Exactly how minutes are written depends upon the type of meeting. There are the so-called constitutional meetings where the minutes become an authoritative record of the proceedings [32]. They provide an accurate historical narrative of the committee's activities. There are also executive minutes where what has been decided upon has to be implemented. They are, therefore, the authoritative documents. Finally, there are progressive minutes that provide a basis for evolving policy which are similar to the progress reports mentioned in Chapter 6.

8.8 Checklist for chairmen

As engineers very often have to convene meetings it is important that they realise they will often be called upon to take over the chairmanship and thus have certain responsibilities.

8.8.1 Chairman's responsibilities

There are certain responsibilities that each chairman of a meeting must take. The ten most important are:

1. Plan and prepare so that the meeting is held under the best conditions.
2. Define the problem(s) and objective(s) to all the meeting members in such a way that their attention is captured and give them a direct interest in the outcome.
3. Help all members to think around the problem and arrive at an agreed conclusion.
4. Keep the discussion alive, to the point and avoid any friction.
5. Encourage all to contribute to their optimum (beware of the commentator, monopoliser and the clam!)
6. Review progress from time to time during the meeting, emphasising areas of agreement and aspects yet to be covered.
7. Where necessary summarise and interpret the opinions put forward so that each viewpoint can be fairly reviewed.
8. State clearly, unambiguously and in summary form, the final outcome of the meeting.
9. Define clearly the action(s) to be taken collectively or by individual members of the meeting.
10. Consolidate the results by follow-up and feedback.

8.8.2 Chairman's qualities

There are certain desirable personal qualities which all chairmen should seek to develop.

1. Patience—he must not be put off by negative reaction—antagonism, cynicism, apathy and shyness must all be dealt with tactfully.
2. Understanding—he should try to be objective and realise that those who take up different positions are not trying to be difficult. They may well be under stress.
3. Firm—there should be no hesitation to take firm and decisive action if the meeting starts to get out of hand.
4. Watchful—he will be ever looking for reactions and keep the

discussion going and moving in the right direction.

5 Analytical—he will constantly be looking for changes in mood in individuals and groups and quieting the discussion appropriately. He will seek to move from obstruction and indifference to co-operation.

6 Receptive—again he must be ready to receive new ideas and new media, etc., to improve his chairmanship.

8.9 Further reading

A Reddin, W. J., 'The incompetent committee', *Business Management* (March 1967).

B Perry, P. J. C., *Hours into Minutes*, BACIE (1966).

C Gordon, W. J. J., *Synectics*, Collier Macmillan (1961).

D Whiting, C. S., *Creative Thinking*, Chapman & Hall (1965).

E Maude, B., *Managing Meetings*, Business Books (1975).

See also References, 27, 33 and 34.

9 Illustrations for talks, lectures, articles and reports

An apt and timely illustration can do much to enliven a dull talk. It can draw attention, focus thinking, aid memory and stir imagination. Also technical articles can be made eye catching by an appropriate diagram or two, which can make understanding easier. Often technical reports need to be illustrated by graphs, schematics and photographs if a proper impact is to be made and an accurate record of the work or project maintained.

Certainly the impact of oral plus visual in communication makes a great difference to the retention of data and information. This effect is known as *synergism*—the simultaneous impact of separate agencies which, together, have a greater total effect than the sum of their individual effects. The key is to arrange the illustrations in the correct sequence so that the combined senses of seeing and hearing are stimulated. This requires careful choice and good timing and placing. In summary it may be said that good illustrations save time, increase interest, help hold attention and increase memory retention.

9.1 Visual aids for talks

There are many types of visual aids that can be used when giving a talk. A few of these have already been indicated in Section 5.7. It is a great advantage to engineers if they can easily prepare their own aids for this avoids delays and high costs. Choosing the correct media and equipment is important. Often the equipment is laid on for the speaker, in which case it is imperative that he makes himself familiar with it before starting the lecture. In professional societies there is generally an operator who should be informed beforehand of when slides and other visual aids are going to be required. Here are a few possible media that can be used when appropriate.

9.1.1 Wall cards

If the speaker wishes to refer from time to time to certain features, a wall chart or card can be very useful. It can be placed on an ordinary blackboard easel to one side of the main lecturing area and pointed to by the speaker when required. Lettering should be bold and preferably done in black ink, but interest and increased impact can be obtained by using coloured poster paint. If words only are shown then one idea per line is a good rule to apply. Diagrams must be drawn with thick lines and anything that is superfluous avoided. Such charts or wall boards are not easily transported and a good strong folder is needed to carry them. Their durability can often be improved by applying strong binding tape [35].

9.1.2 Flannel and magnetic boards

These are sometimes useful where pictures can be built up as the talk proceeds. As each fact is presented so a picture or figure is added. Suitable illustrations from magazines or engineering journals can be cut out and felt pads placed on the back of them. The picture can easily be changed or rearranged rapidly as the talk proceeds. The arrangement can also be made fairly portable.

Magnetic materials can be similarly used on suitable magnetic backing boards and are useful if display panels have to be shown.

9.1.3 Models, maps and drawings

Simple models can be used to great effect when giving a talk and can often be made out of cardboard and pins. However, unless closed-circuit television is used they have to be large if there are many in the audience.

Maps and drawings probably need special preparation if they are to be read with any ease by the audience. Engineering prints are not good material for this reason. One way over the difficulty is to paste a drawing onto some card of suitable thickness and pass it around the audience. This must be done with skill, however, for it is all too easy to lose attention when the drawing is being passed around.

The above aids supplement the normal blackboard chalk and talk method. In using them for recording words, figures and diagrams, the visual presentation grows and develops before the audience. But practice is required even in using the blackboard effectively. Also the speaker, even in academic institutions, is too often found talking at the board and not to the audience. It is sometimes useful to take a photograph (by Polaroid camera) of your blackboard efforts and then analyse them to see if they present the kind of relationships, sequences and constructions you require.

9.1.4 Projectors and slides

These are mainly two types, the normal slide type which may take 2″ × 2″ or larger slides; or overhead projectors using specially prepared transparencies. The advantages of the latter are over-whelming when used properly. The normal slide can be valuable for showing equipment and details of experiments, since it is very difficult to obtain sufficient clarity from a transparency for overhead projectors produced from a photograph. (Conventional photographic negatives can be used but are costly to produce.)

When using slide projectors, lecturers should try to have the presentation under their own command by using a magazine that can be inched forwards or backwards by remote operation under their control.

Overhead projectors are now available in portable form using the Fresnel lens. Top illumination tends to get warm and is best used with a dimmer otherwise the life of the lamp is reduced. Bottom illumination allows a fan to be installed for cooling the lamp and also gives a certain latitude for placing the transparencies. With top illumination the transparency must be absolutely flat on the reflection mirror if all the illustration is to be in focus. The formation of double images is thus avoided. However, the variety of ways in which an overhead projector can be used make it a very powerful medium when giving a talk. By using overlays a picture can be built up as the speaker proceeds. By covering parts of the transparency a gradual unfolding of a picture or set of ideas can be achieved. Again, it is possible to draw with felt-tip pens on transparencies or, where a scroll attachment is fitted, the transparency may be placed under the acetate scroll and marks made with colured felt pen on the scroll itself. In this way it is possible to use the overhead projector like a blackboard. Care should be taken to use an appropriate felt tip pen which leaves marks which are water soluble, so that the transparency or scroll can be cleaned for re-use [30].

A certain amount of practice is required to use the overhead projector to the best advantage, and lecturers need to practise producing their own transparencies.

9.2 Key points for designing visuals

Bernard Shaw is reputed to have remarked about the sign 'Fresh fish sold here', that it was four words too long. The fish is unlikely to be rotten, it is not being given away, it won't be sold anywhere else, and you can tell it is fish by the smell. So why have a sign? All visual designing should seek to acquire Shaw's intuitive contempt for the superfluous. Apply value analysis to your designs. Briefly they should

1 Be bold and big.
2 Have key words drawn 6–8 per line.
3 Keep maximum number of main lines per visual to 6–7.
4 Try to get across one point at a time.
5 Keep it simple.

So do not overload your visuals and to avoid doing this it may be necessary to break one down into several separate transparencies. (30, 35).

It may be that material for your visuals is already complete such as a picture from a magazine, a book illustration, or a line diagram from a report. If you do use such ready-made material make sure that any superfluous lines or features are blanked out and that the printing is large enough. Generally it is best to cut out with scissors and mount the picture and then add the titles and words necessary to make the visual clear.

Some further points which are important to keep in mind when designing visuals are:

1 Seek to present your information in a logical form so that on presentation you can build up point by point with overhead transparencies; this can be done using overlays or uncovering the transparency line by line.
2 The lettering should be bold and at least 6 mm high. Typewriters with bulletin and speech-size type are quite suitable. Alternatively pressure-sensitive lettering sheets may be used.
3 Do not have large areas of your transparency in solid black.
4 Colour can add to a presentation and give greater impact. Colour may be added easily with felt-tipped pens. Alternatively colour transparency sheets can be used to overlay the master drawing.

When producing slides for overhead projectors the thermographic process can be used to produce good quality transparencies from

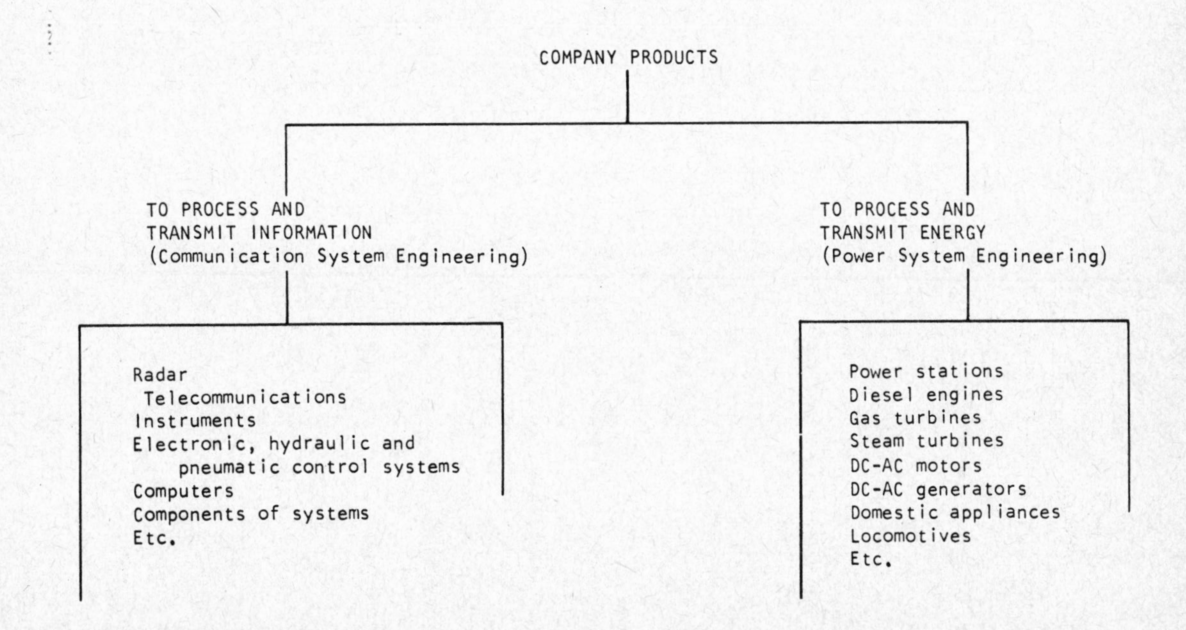

Figure 19 A good transparency of written material

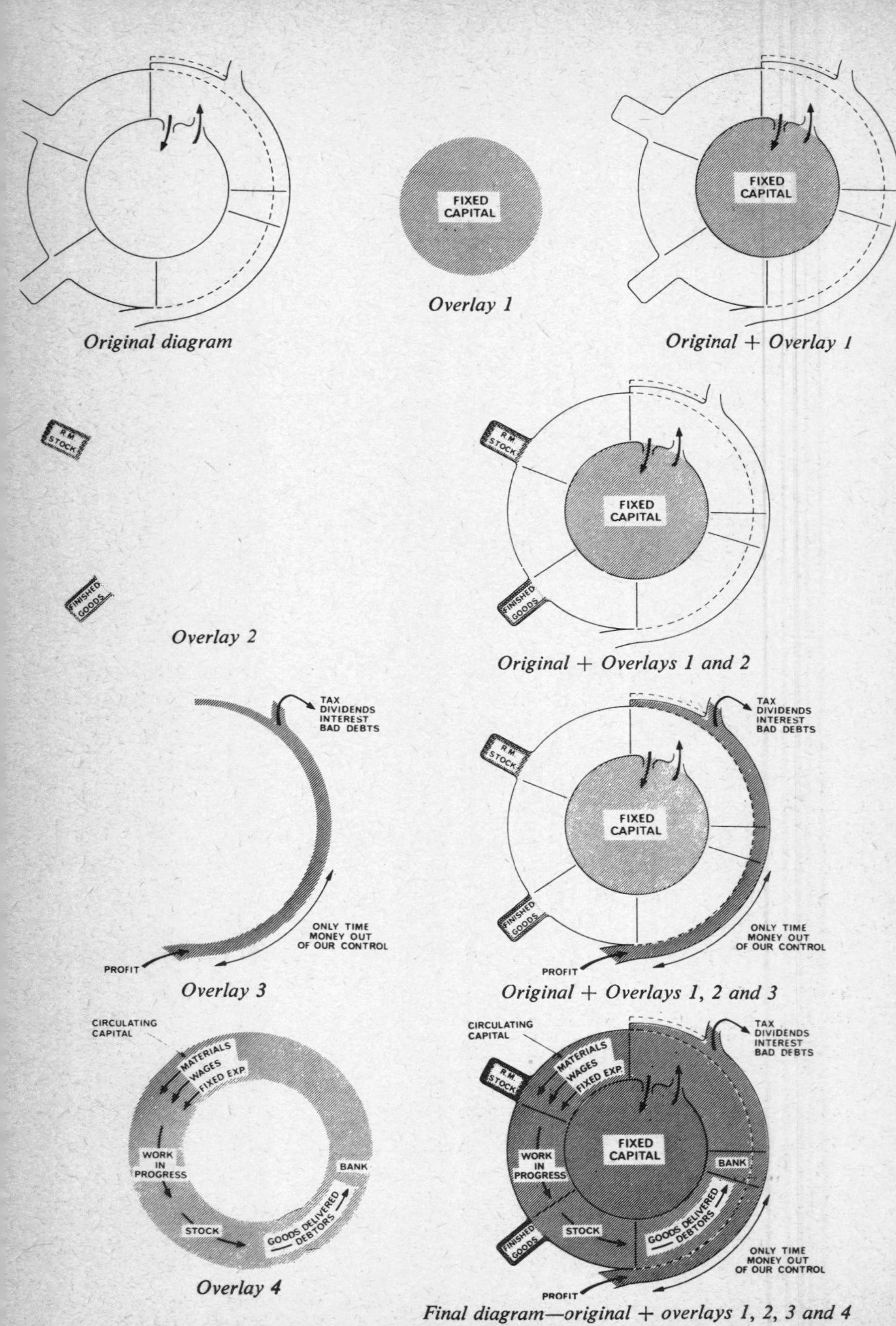

Figure 20 A good transparency of a diagram built up using overlays

felt-tip-drawn diagrams. For good copying there must be a sufficient density of line with sufficient quantity of carbon or metal base: indian ink or new typewriter ribbon. Coloured inks or ball-point pen lines will not reproduce well; electric typewriter ribbons are generally satisfactory. When done with felt pen the drawings have to be 'Xeroxed' first and the prints used for thermographic processing to produce the slide.

When using an infra-red process it may be a good idea to rough out your design in red pencil on blue squared paper and then, when satisfied with the design fill in with black pen or pencil. The thermographic process will not copy red or blue. A typical 'hard copy' of a good transparency of written material can be seen in Fig. 19, whereas Fig. 20 shows a diagram being built up from a basic idea. The additional overlays enable the complete diagram to be developed piece by piece giving the audience time to digest each point.

9.3 Articles and reports

Since a picture is taken in simultaneously 'in one go' and words sequentially, a good line diagram or even a cartoon can be useful in amplifying and clarifying main points in a message. Readers are often attracted to particular articles because of cartoons. A typical example of two by A. E. Beard which were used to illustrate one my own *CME* articles is shown in Fig. 21. It is best to get an expert to do these and generally leave him to choose the incidents to be illustrated. But the editor may well have to say how big and what shape the cartoon or illustration should be.

Confucius said: 'A picture is worth a thousand words', but one sketch may be worth three orthographic views, and one sentence can be worth a thousand dotted lines. It depends what, as a writer, you are trying to show. A good rule to follow is: the illustration must be necessary, it must be clear and it must fit.

An article for the technical press of 6,000 words could take two or three illustrations, one of which might be a photograph. Any photographs used should be on glossy paper and at least 12 by 8 cm. These help to break up the text and make reading more attractive providing they are relevant.

Any graphs must be clear with the axes clearly marked as in Fig. 22. Tabular matter is costly to produce and should be kept to a minimum.

"An attitude in this country that good managers are born not made"

" - the sit next to Nellie attitude - "

Figure 21 Cartoons—the power of illustration

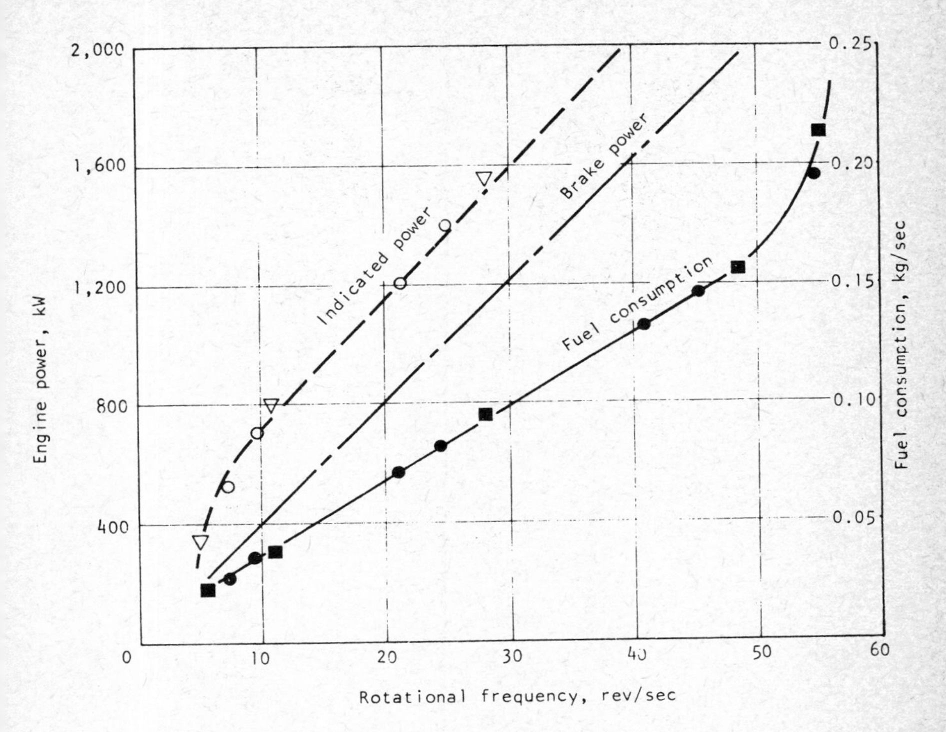

Figure 22 An example of a good graph

For this reason sets of figures should be presented in bar chart form (see Fig. 23).

Various charting systems can be used to great effect. They are, of course, related to graphs. Basically graphs show relationships between things by using two axes, whereas charts use only one. The basic types are shown in Fig. 24 but there are many variations. The art is in choosing the correct variation for your purpose. The simple bar chart shown in Fig. 23 has six variations.

Variation 1—The bars are divided into components to show comparisons within the bar.

Variation 2—This brings together two sets of bars for direct comparison.

Variation 3—This shows an added set of totals for comparison with an original set.

Variation 4—This also divides the bar into components but slides them until the division points align. The dividing line becomes the base.

Variation 5—Also called a deviation chart and is designed to show up differences rather than totals.

Variation 6—This is the 'range' chart as used for depicting engineering tolerances, etc.

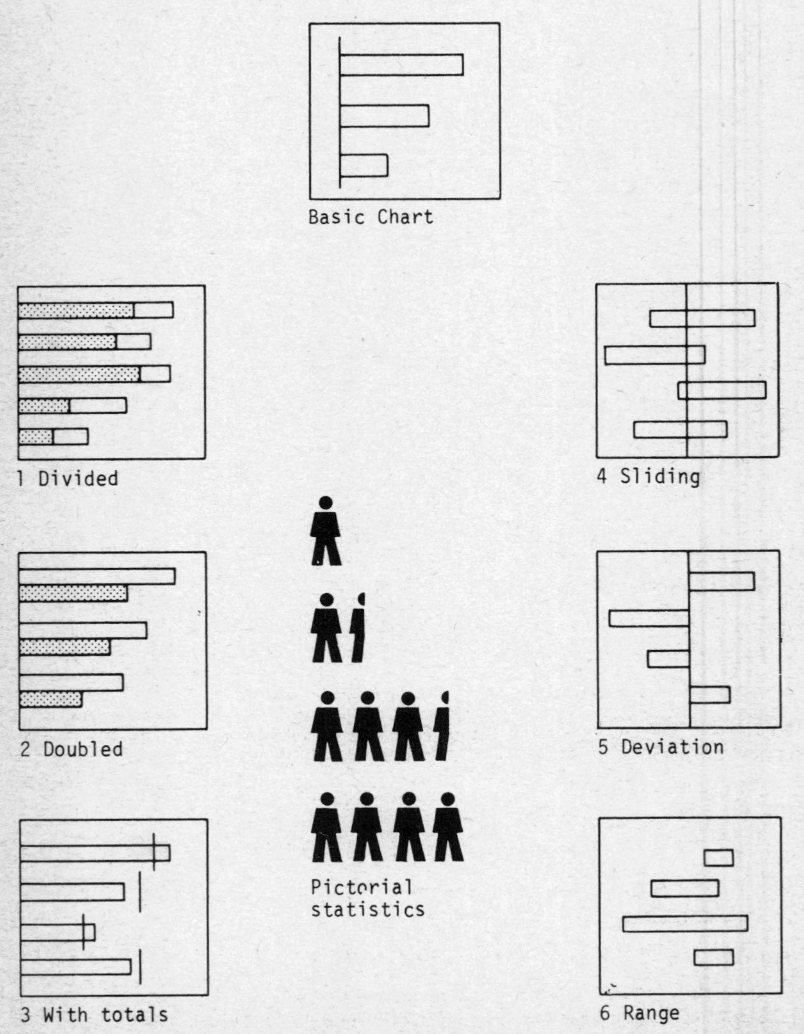

Figure 23 Examples of bar charting

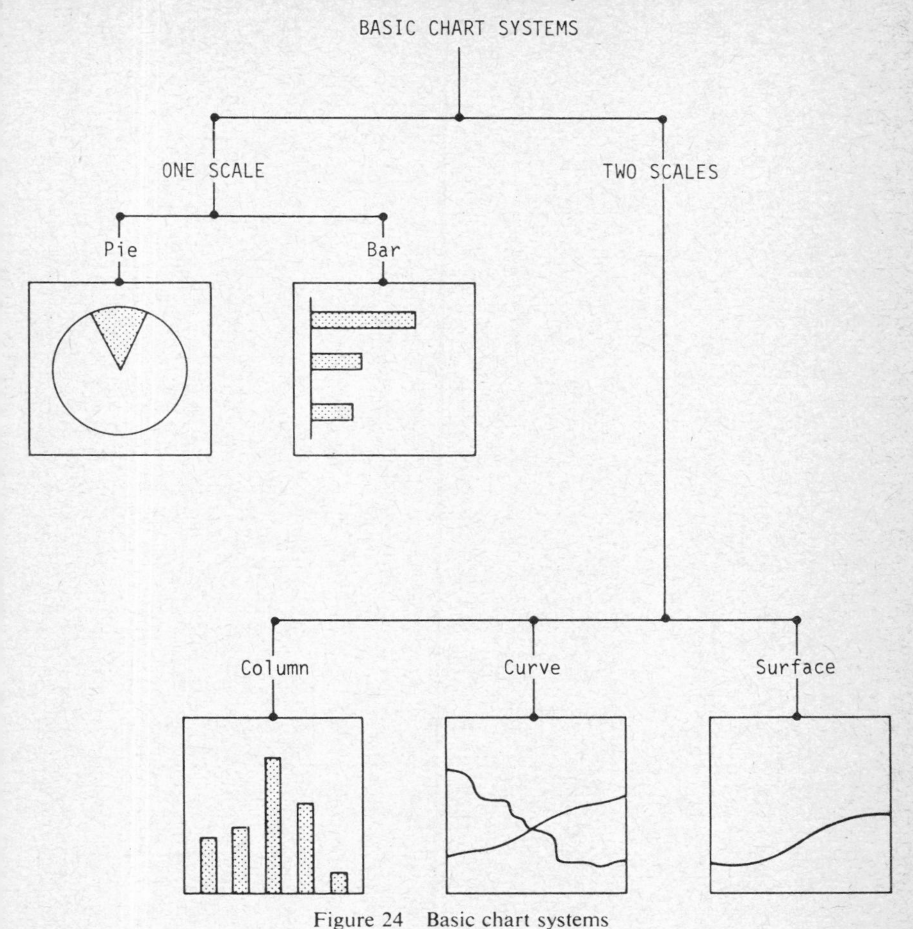

Figure 24 Basic chart systems

9.3.1 *Mathematical formulae and symbols*

Mathematics should be kept to a minimum. Necessary equations should be included in the text each in a line of its own, and they must be numbered consecutively, e.g. (1), (2), and so on; routine mathematics should be omitted or given as an appendix. Mathematical formulae will generally be set in italics. The solidus is preferred for any fractions in the text or in illustrations [for example $(x + 2q)/(p + r)$], as they can be written in one line. Likewise, some journals prefer the square root to be written without a bar thus, $\surd x$ or $\surd(x + 2y)$—many prefer x^{1}/y to $\surd x/y$, especially as the latter is ambiguous. Similarly horizontal bars should be avoided as they require hand setting so write $n!$ not $\underline{|n}$.

The use of dots and bars to create new symbols such as $\dot{a}$, $\ddot{b}$, $\dot{\bar{e}}$, $\bar{f}$, $\bar{\bar{h}}$, should be avoided if possible because they are difficult for the printer and involve double spacing of the whole paragraph when used in the text. The asterisk (*) should not be used with a symbol unless referring to a footnote.

Only simple symbols and formulae should be typewritten. All others should be written clearly in ink and Greek letters should be indicated, preferably spelt out in the margin on first use, e.g. Greek mu. A clear distinction should be drawn between capital and small letters, especially in suffixes and prefixes and between inferiors, superiors and primes. Letters and symbols which are specially liable to confusion are alpha and 'a', kappa and 'k', mu and 'u', nu and 'v', eta and 'n', the letter 'l' and the numeral 1, and the letter 'o' and the zero 0. All symbols used should be defined at the beginning of the mathematical section or appendix.

Two-level and complicated suffixes, e.g. V_{s_a} and exponents should be avoided as far as possible as they are inconvenient for the printer. Every scientific journal has special preferences and authors should obtain details from editors before embarking on their final texts.

It is important to ensure that any diagrams, graphs or other illustrations contain the same symbol types as the text. In graphs and diagrams presenting experimental results, it has been found that those

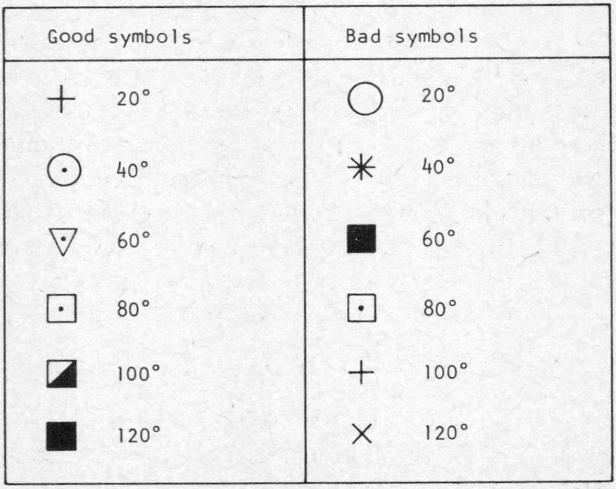

Good symbols		Bad symbols	
+	20°	○	20°
⊙	40°	✳	40°
▽ (dotted)	60°	■	60°
⊡	80°	⊡	80°
◪	100°	+	100°
■	120°	×	120°

Figure 25 Symbols for plotting experimental results on graphs

shown on the right-hand side of Fig. 25 are bad because if the two points representing results for 100° and 120° become superimposed they may be misread as one result for 40°, thus leading to ambiguity. A further objection to the of symbols as shown on the right of the table is that the eye is most impressed by the heavier symbols and the scale of impression on the eye caused by these reads 100°, 120°, 20°, 40°, 80°, and 60° to most observers. Where possible the scale of impression on the eye is drawn through the sequence 20°, 40°, 60°, 80°, 100°, 120° in the proper order when the symbols shown on the left are used [A, B].

Sometimes pictorial symbols may be used to give greater impact in presentation as an alternative to ordinary bar-charting (see Fig. 23).

When depicting flow charts special symbols should be used (see Fig. 26). For computer work these have been standardised and a typical example is shown in Fig. 27. This represents the writing process and should be compared with Fig. 13 and Table 2.

9.3.2 Metrication and technical writing

The Royal Society Conference of Editors, after considering the role that scientific and technical publications can play in the government's policy of promoting the general acceptance of the metric system, has made two main recommendations:

1 That the SI (Système International) should be adopted for all scientific and technical journals.
2 That the changeover should be effected as soon as possible.

It is important that technical writers seize the opportunity of playing a crucial role in helping to end the confusion and waste (both mental and material) resulting from the present multiplicity of units [D].

The rate of a changeover towards complete metrication will obviously vary from firm to firm and between technical journals depending upon the subject covered and the extent to which the metric system already operates. In certain firms the writer of technical literature may need to proceed to the target by the following route:

Non-metric (SI) → SI (non-metric) → SI
(i) (ii) (iii)

In certain industries full metrication will have to wait upon the installation of metric machinery and equipment but in many engineering publications the change over to SI units can be achieved in one step.

Symbol	Meaning	Symbol	Meaning
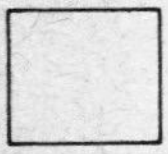	Process		On-line storage
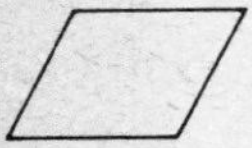	Input-output	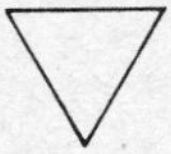	Off-line storage
	Decision		Document
	Terminal		Punched card
	Connector	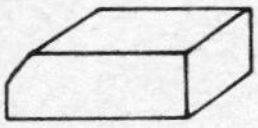	Deck of cards
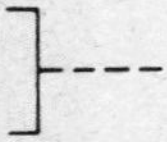	Comment annotation		Punched tape
	Manual operation		Magnetic tape
	Merge		Magnetic drum
	Extract		Magnetic disc
	Collate		Graph plotter
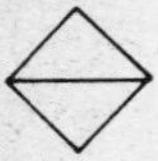	Sort		Visual display

Figure 26 Program and system flow chart symbols

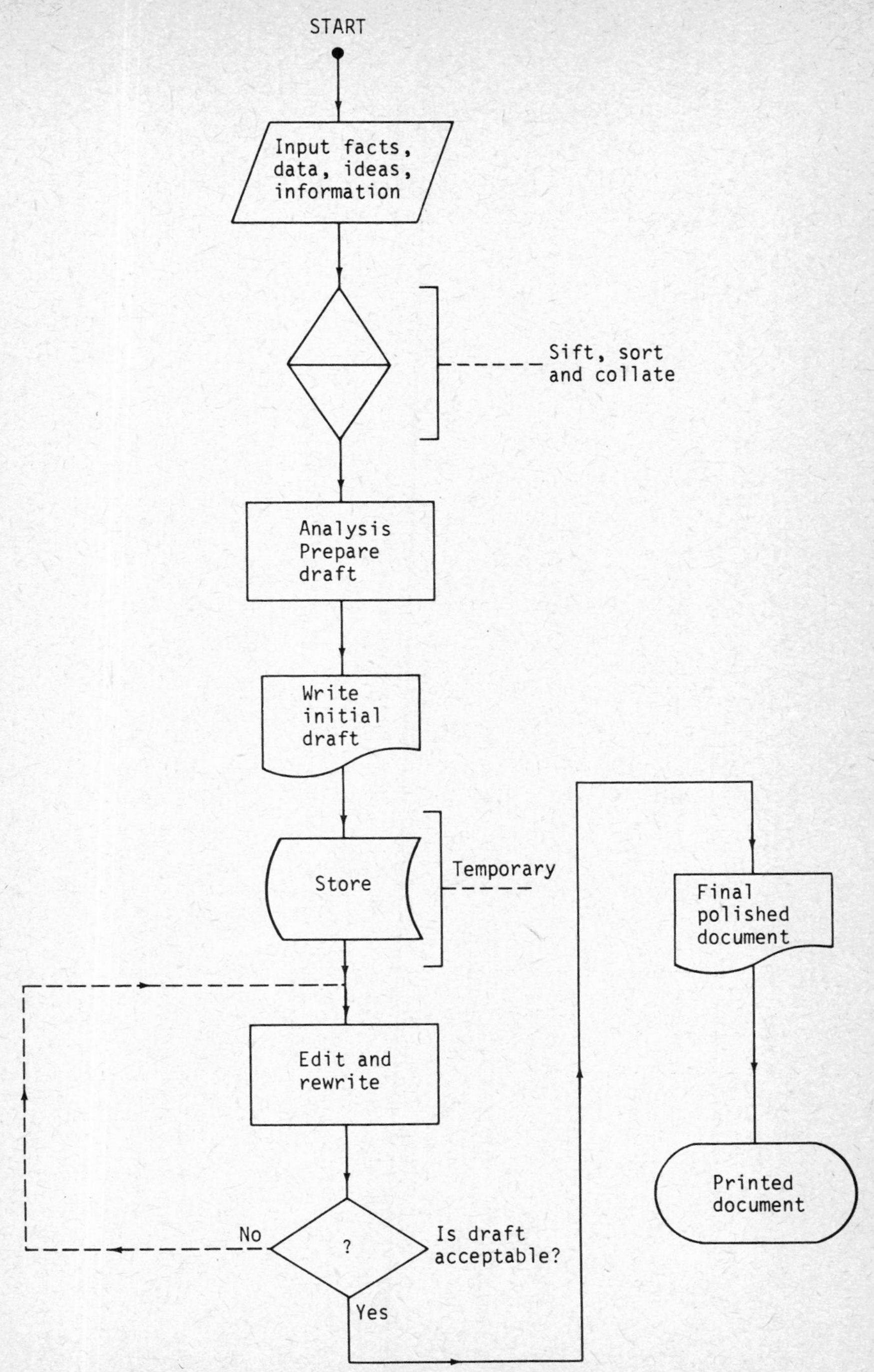

Figure 27 The writing process illustrated by using flow chart symbols

For reference purposes the basic and supplementary SI units and the derived SI units are given in Appendix 14.

9.4 Further reading

A Barrett, A., 'Linking research and design', *Aslib Proceedings, 14*(12), (1967).

B Rhodes, F. H., *Technical Report Writing*, Chapter 8, McGraw-Hill (1961).

C Lewis, B. N., *Flow Charts, Logical Trees and Algorithms for Rules and Regulations*, CAS Occasional papers, No. 2.

D Mayhew, Y. R., 'The use of SI units in IMechE publications', *CME* (October 1973).

See also References 35, 36 and 37,

10 Handbooks and manuals

Unfortunately many good products are delivered with inferior servicing instructions and technical handbooks. Sometimes, at the handover stage, a customer is given no proper manual. This may cause operator and maintenance training to be delayed until there is virtually no time to do any before the equipment is commissioned. This condition leads to a panic situation where something is hurriedly compiled to enable operators to get a system working. Inevitably mistakes are made and some of these can prove costly. If the product is to be used abroad and is part of an export order the situation can be even worse. Management must ensure that such documents are written in phase with the design and development process. On military contracts it is customary to award separate contracts for such work and where this is done it is generally possible to start compiling at the design formulation stage.

Clearly written users' manuals are a must for many domestic products. Users need to know how best to use their equipment and maintain it in good working order. In addition workshop manuals or other servicing publications are required by sales and distribution agents. Too often these documents are poorly written and presented. Many manuals do not specify the type of tools to be used for various operations and frequently fail to mention the sizes of nuts and bolts which have to be removed. Quite frequently such service manuals have to go to operators in the developing countries. There are many pitfalls in technical writing for these purposes. For example, writers often refer to one item using different words—grub screw, lock screw, collet screw etc. Scrupulous care must be taken to use the correct term and maintain consistency throughout the text for documents to be used abroad otherwise translation becomes difficult.

Another fault is finding assembly and test procedures alluded to but no means of carrying them out given, e.g. 'after assembly—check alignment'—this bold statement is all that is given and no reference is made to jigs and fixtures or how measurements are to be made.

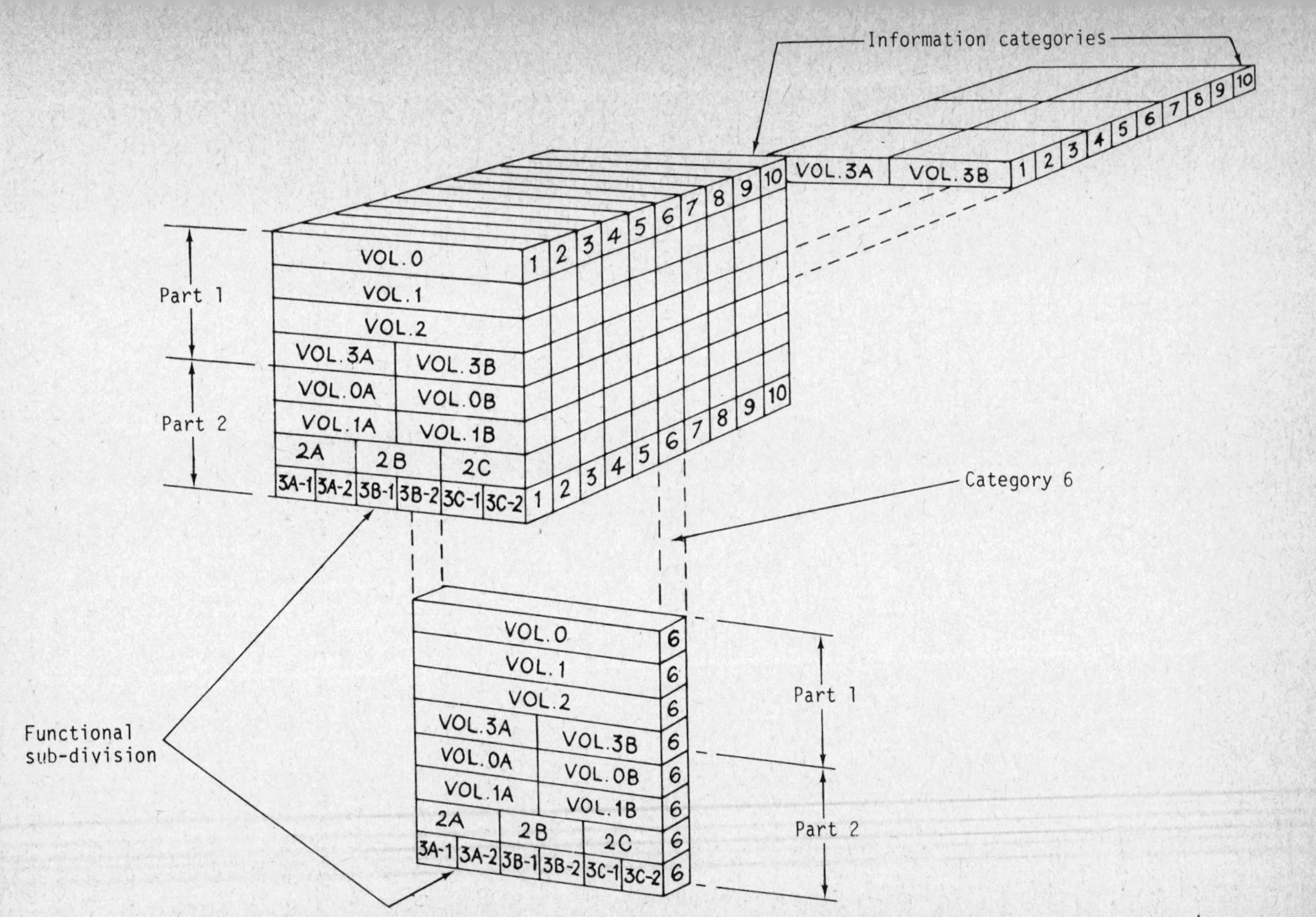

Figure 28 Manual organisation by category: contains information relating to all functional subdivisions—category→part→volume

10.1 The manuals compiler

This term is used advisedly because the man or woman who prepares manuals of one sort or another is working in a tighter discipline than an author or editor of papers or articles. It is his, or her, job to ensure that the product or system can be maintained and operated with maximum efficiency. They must therefore include certain minimum information. They must tell the operator what the equipment does and where it is. Manuals are not additional sales literature: blurb such as: 'This gubbins has been designed with great care and if used properly will last for ever' must be cut out. Manuals can be divided into two categories: instruction manuals and reference manuals.

The compiler will generally group his information along these lines:

1 *General information* containing an introduction, a brief description and technical summary.
2 *Instructions for use* containing installation details, setting-up arrangements and operation details.
3 *Maintenance* containing a detailed description, routine maintenance overhaul and repair and parts list, circuit diagrams etc.

This is one way of organising a handbook: there are others. A reference manual will embrace all three sections and an instruction manual will generally be confined to (1) and (2). While this approach can apply to all types of equipment whatever its size and complexity, it is important that the manual is so structured that functions are separated into sub-divisions.

Looked at in three dimensions a desirable method of organisation can be seen in Fig. 28. In any system the compiler has to consider spatial envelopes as well as functional entities. This is known as the concept of separately maintainable sub-systems illustrated in Fig. 29. Such a sub-assembly breakdown can then be organised into various levels similar to Fig. 30. Here a sound recording system has been broken down and placed on four levels.

Having set out a logical structure the compiler has to organise the technical detail in each slot. He therefore has to be familiar with all the technical design features if he is to produce worthwhile work. This has led to the use of the term 'design disclosure'. Complex equipment will necessitate a three-pronged attack to get desirable documentation. Design will be concerned with both the specification and maintenance

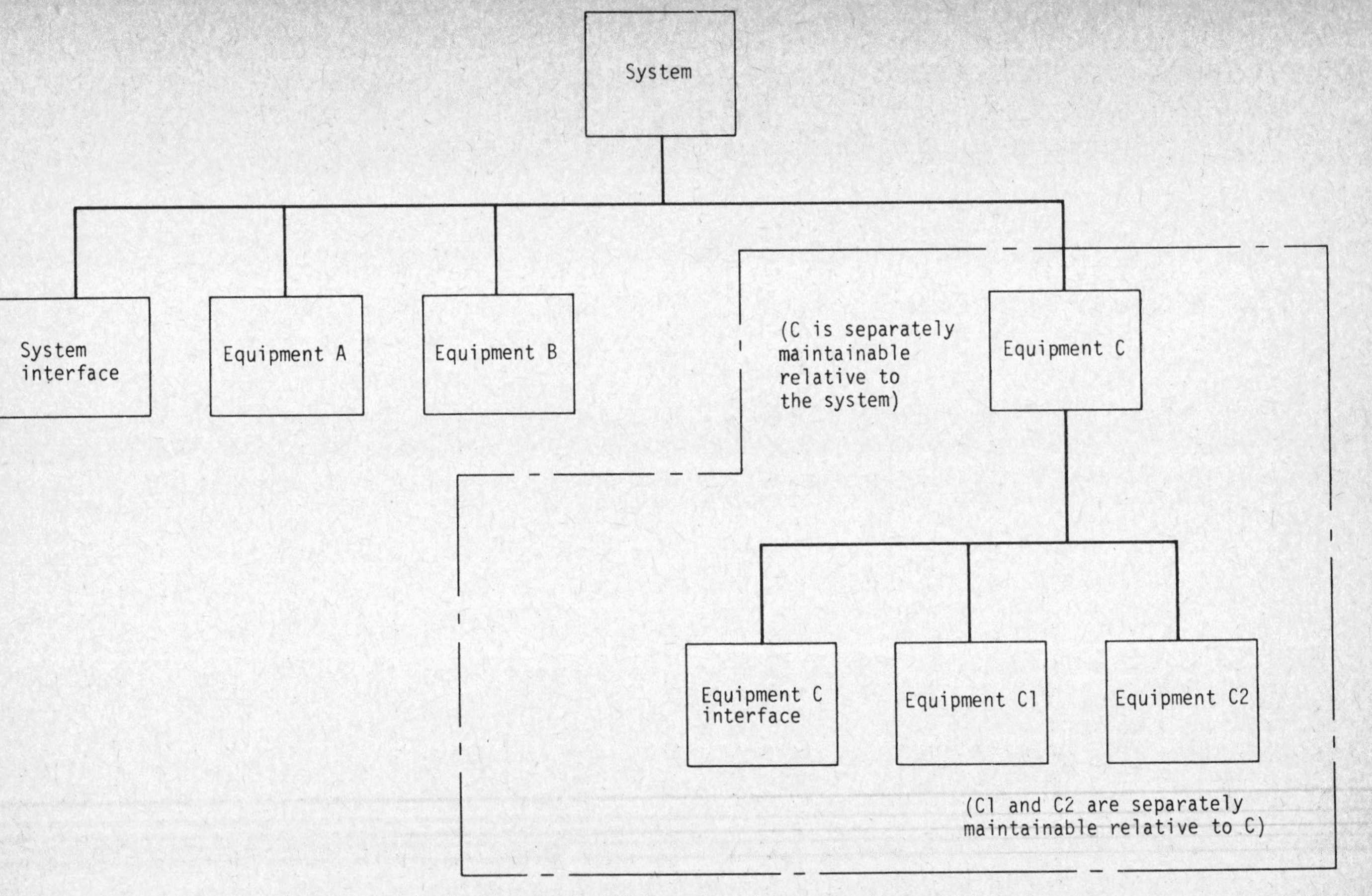

Figure 29 System and subsystem breakdown

policy, for this will affect design strategy and detail. But production will also be involved as regards spares, levels of stocks and production of tool kits, etc. The user will also wish to be consulted on these features since he will generally be interested in minimising down time in the event of some fault occurring.

10.2 Fault-finding manuals

Great strides have been made in improving maintenance manuals by the introduction of maintenance dependency charts. Systems have been developed by the armed forces and by industrial firms for this purpose. To produce such charts a special design documentation system is set up so that there can be a clear understanding of the development at all stages.

A design documentation system (DDS) also provides a means of design control not often achieved by other systems, particularly in the area of equipment maintainability. A functionally identified maintenance system (FIMS) greatly simplifies and speeds up maintenance and allows a much fuller use of semi-skilled personnel [37].

10.2.1 Design documentation system

Where such a system has been installed for a large contract it is quite common for a DDS co-ordination team to work under a project manager. Such teams collect and collate all project information and keep it up to date, as well as publishing progress reports, etc. Special formats are produced to cover the levels as indicated in Fig. 30. These correspond to the master level—covering the overall system—in this particular case the sound recording and reproducing system. The intermediate level covers the chains of related functions, assemblies and components and so on. It is possible to sub-divide at any level but the important feature is that there must be a continuous feedback between the various levels.

The documentation required from the designers at each level varies with the type or work and kind of project but generally consists of:

A functional block diagram
A functional text for each block.
A design outline
Supplementary information

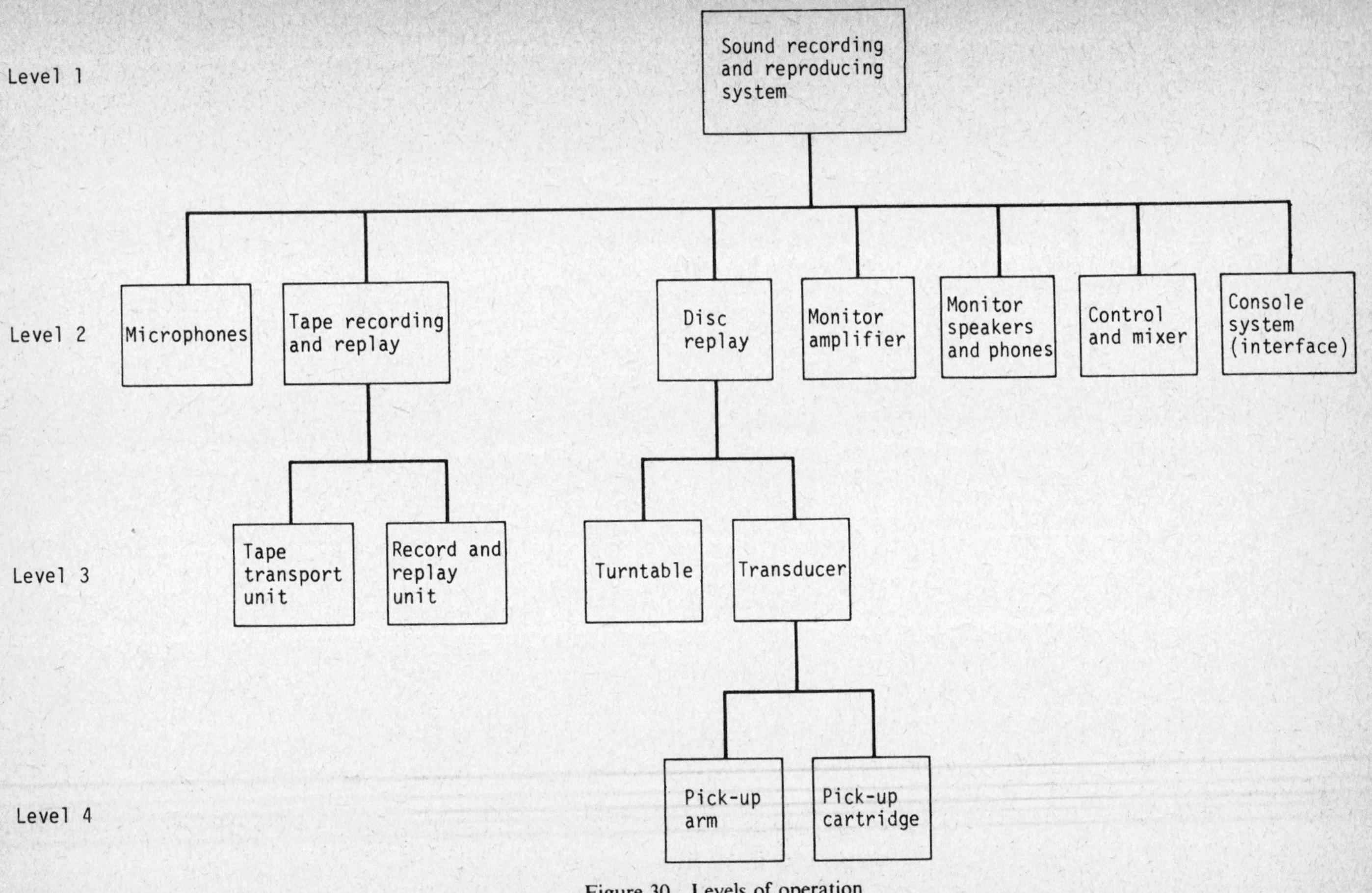

Figure 30 Levels of operation

The functional block diagram defines the various system functions and shows the information flow between functions, defines main signal flow and feedback paths, defines areas of hardware and indicates all service requirements, such as power and cooling. The text for each block describes the function and general theory of operation. Important design data such as 'mean time between failures' and 'estimated time to repair' are also quoted. The design outline will contain details such as: the logic dependency level at each stage, the relationships between functional entities and detailed specifications proving input/output interface compatibility. An assessment should also be included of any maintenance requirements. Sometimes an assessment of penalties incurred by introducing additional stages is also given. Supporting theory, parts lists, mechanical layout test procedures, etc., are included in supplementary information.

10.2.2 Functionally identified maintenance system

This is one system that enables logical, detailed fault-finding to a degree well beyond that which a semi-skilled technician could normally provide. It also ensures that the equipment itself is carefully analysed for maintenability as design proceeds and before it is handed over to the customer. The heart of the system is the maintenance dependency chart. Where these charts are properly drawn a maintainer will be taken through a chain step by step, full information being provided to allow him to locate the function, check its input and output dependencies, etc. On many such charts the test data is printed to one side of the chart and on the other side the test procedures are outlined. A typical layout of a maintenance dependency chart can be seen in Fig. 31. Here the fault isolation procedure is as follows: the maintainer notes the faulty equipment indicators and selects the appropriate fault dependency chart. He then proceeds to establish the first bad indication and the last good indication of the dependency structure. The dependency structure between these two events then contains the faulty structure. The maintainer selects the mid-point of the faulty element and chooses an easy access point to narrow down the area of concern. If the event chosen indicates 'Go', then the fault lies downstream, if the indicator shows 'No go' the fault lies upstream on the dependency chart. The procedure is continued moving appropriately until a single dependency line is isolated where the input event is good and the output bad.

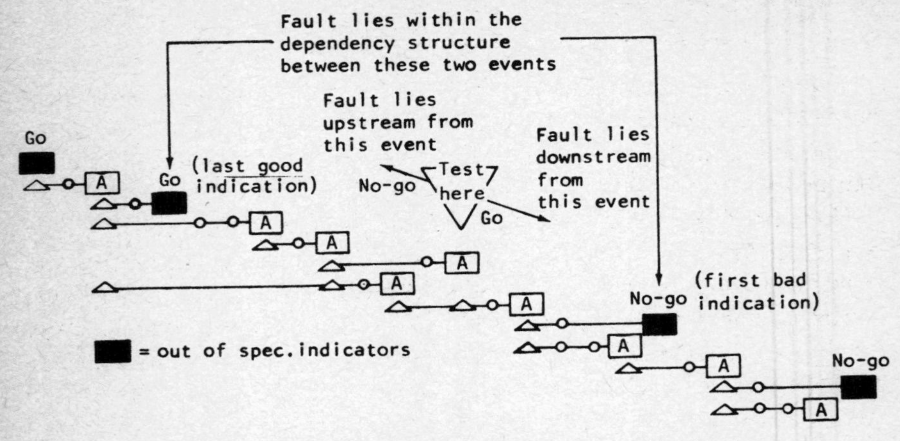

Figure 31 Maintenance dependency chart for trouble shooting

The manuals compiler must, therefore, make sure that such dependency charts are prepared and built-up during the design process when such a system as this is used.

10.2.3 The use of programmed texts as manuals

In addition to maintenance manuals, special manuals are sometimes required to instruct sales engineers and overseas agents in the theory of the latest types of equipment being produced by their companies. It is often very expensive to bring all these people home for special training, and it has been found desirable to produce 'scrambled' or 'fractured' textbook types of manuals for this purpose. The amount of information that has to be absorbed can be large, and a sales engineer's time is often precious, so such programmed texts can be invaluable in cutting costs.

Here again the requirement is for clarity of presentation using adequate diagrams built up as the reader goes through the text. The basics of such a programmed instruction text are that the material is presented so that:

1 It follows a logical sequence.
2 Is presented in small steps.

3 The reader is tested after each step.
4 Confirmation of the correct answer is given immediately.

From such a presentation it follows that service engineers can proceed with their learning at their own pace and difficult material can be learnt more easily and psychological barriers (such as age or status) can be overcome.

Considerable skill is required in writing such books since it is a matter of knowing what to exclude as well as include. An essential in communicating information to other people by this method is for the writer to know what they do not know, and to try to convey at every level, not just one level, the essential features the writer wants them to acquire at the end in terms of behaviour. Such manuals have been used very successfully by several firms and have been validated over a period of time. They have shown considerable cost savings over normal in-plant training methods.

10.3 Special technical literature and instructions

Sometimes special technical publications have to be produced for operating machines, test beds or standby equipment. Many of these can involve the presentation of complex interrelated rules, give much trouble and often lead to very involved technical literature. Others may be covered by verbal instructions which describe operations to be performed in a single sentence. For example how to connect up to a power supply might be covered by a sentence like, 'Connect brown lead to the terminal marked positive (+); the green/yellow lead to the terminal earth (E)'. This is simple to follow and no problems arise provided there is no sequence problem.

But if the instructions have to be interrelated it is not so easy and such trouble occurs in many instruction manuals and also in hire purchase agreements and other legal documents.

One of the best examples is in the official leaflet N195 published by the Ministry of Social Security in 1967 in the U.K. This tried to explain the benefits available to women whose marriages have ended in divorce when they were over 60 years old!

> 'If all conditions are satisfied, you can qualify on your former husband's contribution record for a pension equal to that which you would have received had he died on the date your marriage

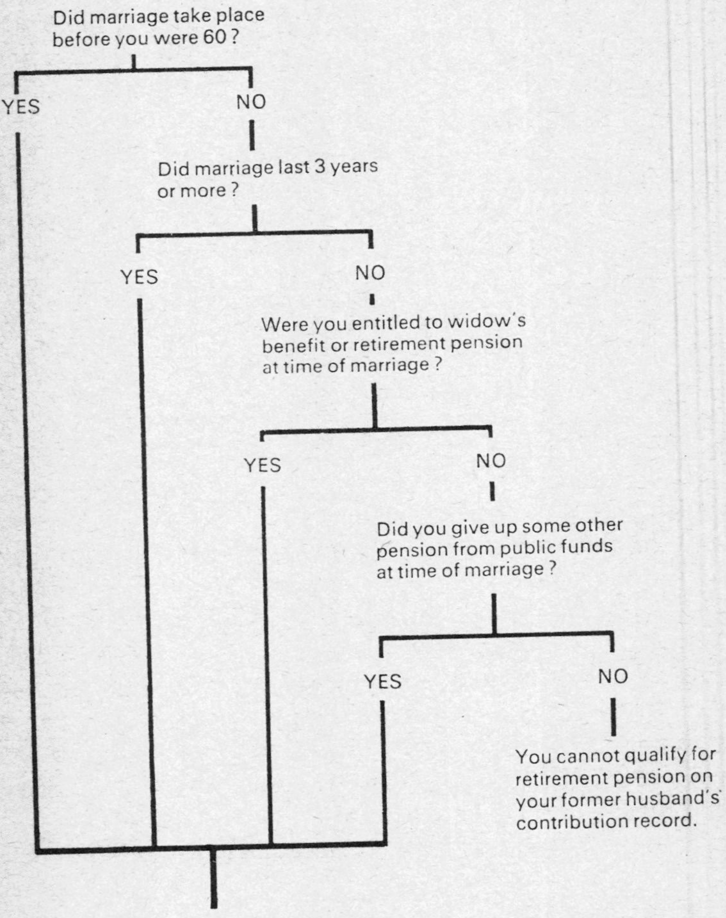

Figure 32 An example of a logic tree

ended, regardless of whether you have, or your former husband has, retired or not. If the marriage took place after you were 60 this rule will help you only if the marriage lasted three years or more or if, when you married, you were entitled to widow's benefit or retirement pension under the National Insurance Scheme or had to give up some other pension paid from public funds'.

A simple logic tree of this confused writing can be seen in Fig. 32 and is obviously a far better means of communication.

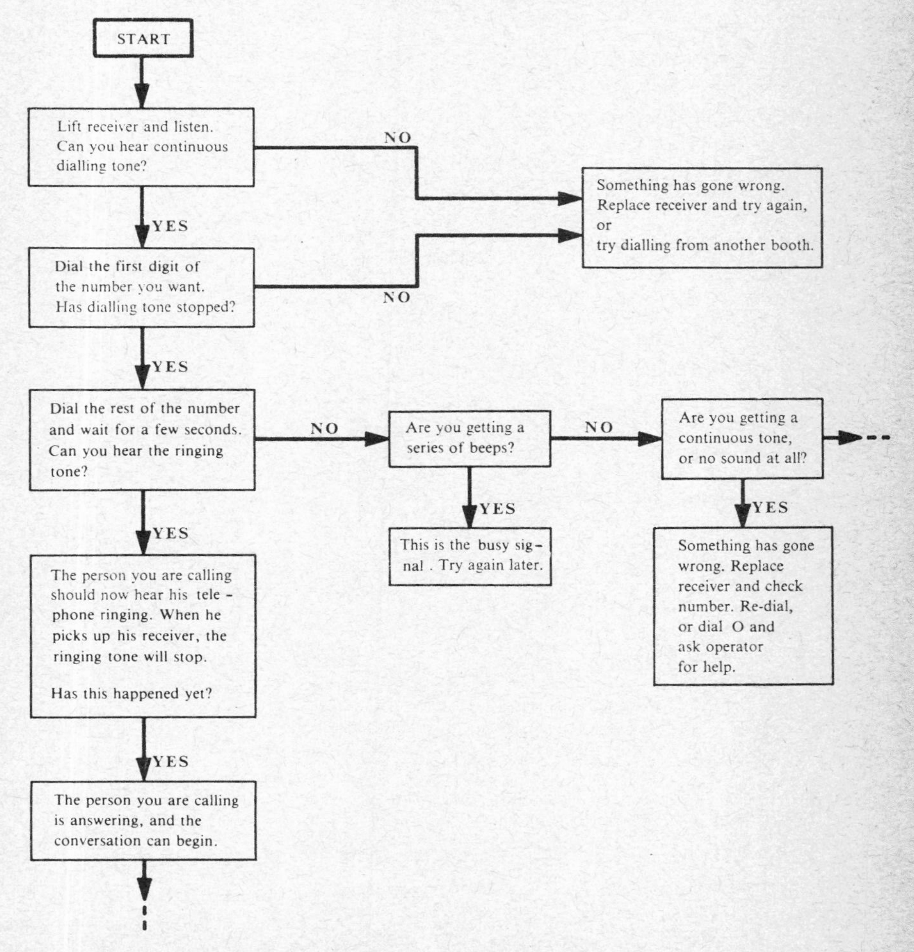

Figure 33 An algorithm for making a local call

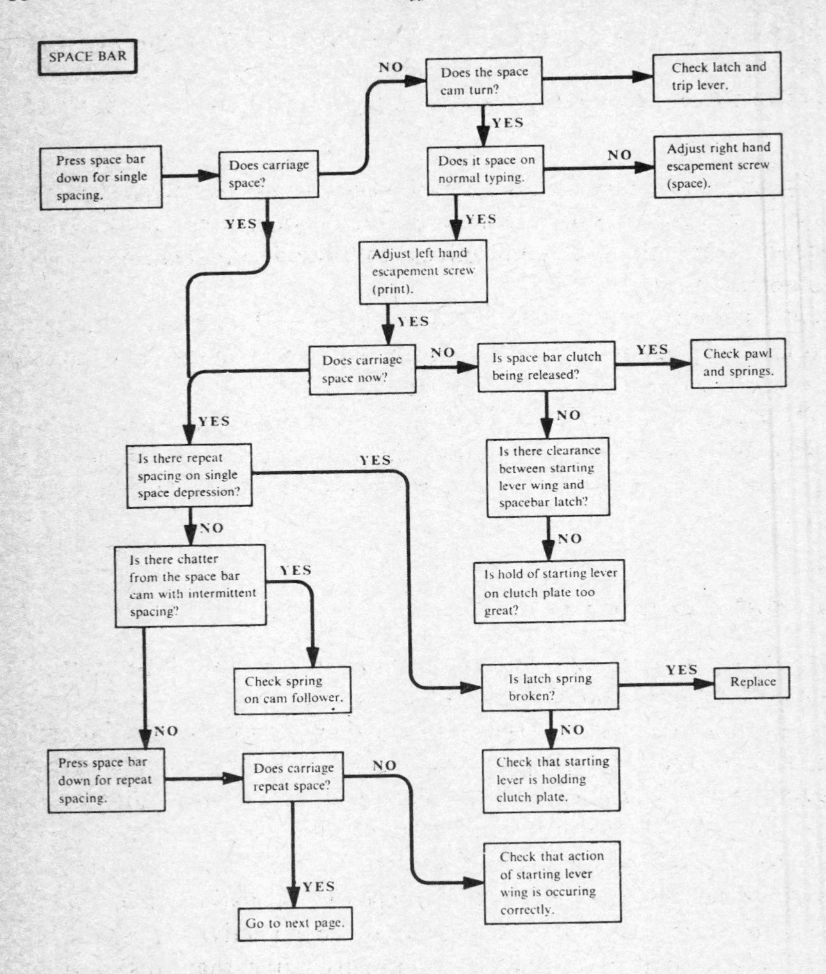

Figure 34 An algorithm for locating faults in an electric typewriter

A typical example more applicable to the technical world would be the use of the telephone. Here in place of involved interrelated statements the operator proceeds in a series of logical stages which are shown diagrammatically in Fig. 33. This method allows a sequence of simple technical statements or questions to be ordered in a logical hierarchy from the most general to the most specific, in such a way that only sentences which are relevant to the particular situation need be read. This method can be well adapted to testing instruction manuals.

Algorithms can also be used to aid locating faults in mechanisms, etc.; an example of one for locating faults in an electric typewriter can be seen in Fig. 34. Such presentations can be used to great effect in technical writing and have three advantages over conventional prose:

1 The user is required to make explicit decisions, usually of a binary type (Yes/No) about a limited series of questions or sentences. This means that all connectives and qualifying clauses can be eliminated.
2 Only the minimum number of decisions need be made by the user. But in normal sequential writing the relevant statements have to be abstracted and checked against the facts.
3 The reader in proceeding down the logic tree or algorithm does not have to remember previous decisions. In continuous prose, it is necessary to remember the implicit decisions that have been made previously.

10.4 Staff requirements

The staff of a technical publications department must be carefully chosen. The actual numbers will, of course, depend upon the activity level of the firm concerned, but basically there will be a need for technical authors whose job it is to write clearly about the various aspects of equipment. They will need to be helped by technical illustrators, whose job it is to produce block diagrams and any logic and circuit diagrams where appropriate, and perspective views of equipment. As previously mentioned their work is to clarify and make explicit certain aspects of equipment. Often, exploded and perspective views will be a powerful aid here and tend to be much better than orthographic engineering drawings. All such illustrations must have the correct 'fit', i.e. they must be fully interpreted in the text nearby. Just as the librettist who is writing the words for an opera has to work in harness with a composer so the technical publications department must have illustrators working with authors in complete harmony. In addition, on large contracts there will be a requirement for specification schedulers and cataloguers, whose job is to fulfil the demands of production departments and to provide servicing personnel with accurate data for replacement of parts, components, etc.

Finally any publications department must have the necessary supporting services to aid the above categories of staff, such as reprographic and photographic experts. All these staff members have to work together as a team. The compiler may be likened to a film script writer who must work in conjunction with a producer, a director and performers to produce a group project.

A constant problem that occurs in a technical publications department is estimating cost and time.

10.4.1 Publications estimating

Technical writing like any other commercial activity needs to be carried out with adequate cost control. Initially this demands setting up standards of effort and work content so that comparisons can be made for budget control, programming and progress assessment.

Generally a specification of the proposed technical publication has to be written. The smallest sub-unit of such a specification is that chapter in the manual. Some chapters will be longer and harder to write and, when estimating, not only technical complexity of the equipment must be taken into account, but the customers' servicing requirements too. These factors are generally reflected in the chapter headings. From these it is possible to define the cost of a standard chapter. Such a standard can be varied according to the types of labour which will be needed and the difficulty of assembling the relevant information. (38).

The standard unit of effort is generally an author/month and consists of an author working for one month with the necessary back-up facilities and personnel. Standards may, therefore, be built-up in terms of chapters per author per month. These may then be used for estimating by applying to the average type of chapter and the average chapter time the appropriate cost units. A typical build-up of time units and service loadings can be seen in Table 4.

Generally speaking a standard chapter consists of about 13 pages of A4 manuscript backed by 8 illustrations which, set into type and blocks condenses to about 4 pages. It has been found statistically that an author requires about 70 per cent of an illustrator's time on a difficult technical chapter.

The big problem with estimating however occurs when amendments have to be made to chapters already written. In R & D contracts this can apply to 30 per cent of all chapters written. In this case it is necessary

Table 4 Basis for estimating writing costs

Aspects		Work contents	Effort				Cost at present rates
Prime units	The chapter	The smallest division of a publication retaining technical completeness	The author-month:	the work of one author plus a ratio of ancillary services over one month			
			Author	Drawing	Clerical	Supervision	
Detailed explanatory definitions of the units	Type 'A'	Reduced editorial and drawing office content	1.00	0.25	0.06	0.17	Costs vary with efficiency and over-heads
	Type 'B'	Normal chapter	1.00	0.50	0.05	0.10	
	Type 'C'	Difficult technical content of designs liable to change and containing a contingency for resultant amendment effort	1.35	0.675	0.078	0.135	

Table 5 Basis for estimating printing costs

Type of service	Units	Output rate	Average quantity per chapter
Typing	Foolscap pages—double spaced	10 pages/hr	13 per draft chapter
Checking	Foolscap pages—double spaced	10 pages/hr	13 per draft chapter
Photography	Tone drawings—bromide prints	2 chapters/day	3 per user handbook chapter
	Line drawings—bromide or Xerox prints	,,	11 per chapter
Justified typing	Justified sheets (includes automatic double typing process on Varityper: Justowriter tape correction and reading	2 sheets/hr (average for all processes)	8 per printed chapter
Lithography	Line standard pages		8 per chapter
	Foldout pages	2 chapters/day	3 per user handbook chapter
	Half tone illustrations		
Printing collating, etc.	2 printed pages = 1 leaf	32 per day	16 pages per printed chapter

to have some rational method of estimating the additional factor. In my experience severe amendment adds about one third to the cost of any chapter.

Having arrived at the total work content in terms of standard chapters the required effort can be determined in terms of author, drawing, clerical and supervisory effort from Table 4. According to the standard chapter chosen, adjustments can be made for more or less difficult technical content.

When estimating the printing and after the author has completed his work the figures in Table 5 have been found to give reasonably reliable time factors and hence appropriate costs.

10.5 Further reading

A Reul, R. I., *Management of the Maintenance Function*, InComTec Manual (1966).

B Jost, H. P., 'Down with maintenance—ten simple thoughts on the complex subject of terotechnology', *Chartered Mechanical Engineer* (April 1972).

C Robertson, A. G., *Quality Control and Reliability*, Pitman Press (1971).

D Bruce, J. G., *Train Formations and Maintenance Requirements*, Railway Division, Institution of Mechanical Engineers (January 1971).

E HMSO, *Working Party on Maintenance Engineering* (April 1970)

F Turner, B. T., 'Design for society—planned design can simplify maintenance', *Design Engineering* (May 1971).

G Parkes, D., *Terotechnology—the changing face of plant engineering* paper presented at the International Plant Engineering Conference, American Institute of Plant Engineers (September 1971).

See also Reference 37.

11 Patents and copyright

In nearly every country the law provides protection for artists, authors and inventors. Scientists and engineers who work in research and development often have to protect their work by taking out 'Letters Patent'. This gives the monopoly right to make, use and sell an invention. To earn this right the engineer or scientist must provide a full and sufficient disclosure of the invention to enable anyone skilled in the art concerned to understand and make the invention. Also for the monopoly to be valid the inventor must actually try to exploit it and not merely use his right to prevent others doing so.

In our free enterprise world the patent system has evolved to encourage the production to new information. But the form it has now taken has been influenced by the desire to make such new information as freely available as is compatible with the main aim of providing an incentive for invention [39].

11.1 Patent writing

Strictly speaking patent writing is a specialist's job, but it does form a branch of technical writing. Many companies employ their own patent agents who deal directly with the Patent Office. Individuals, such as a designer, may however patent a mechanism or a special type of gearbox if they wish to do so but they would normally be advised to employ a professional patent agent. This is because the drafting of the patent specification requires not only considerable technical knowledge but also legal understanding if it is to succeed with the Patent Office examiner.

An invention was defined by the *Patents Act, 1949* as 'any manner of new manufacture' or 'any new method or process of testing applicable to the improvement or control of manufacture'. Clearly this excludes any basic scientific principles or mathematical formulae.

Once a patent has been granted it can be kept in force for 16 years provided the necessary renewal fees are paid every fourth year. These fees rise from £6 in the fifth year to £30 in the sixteenth year.

The engineer can give considerable assistance to his patent agent by setting down in his own words exactly what his invention comprises, an exercise which may well tax his writing ability. He must describe in words aided by diagrams a claim that something is entirely new and original. In doing this he will undoubtedly be rewarded for as with specification writing, discussed in Chapter 2, the very exercise will help to clarify his thoughts. In addition he will learn to be able to read patent specifications which may well be a fruitful field for stimulating fresh thinking. When a patent search is carried out it may well be that other existing specifications will throw additional light on his own invention. Patent specifications can be a very useful source of information to research and development engineers as well as designers.

The most important thing is for the inventor to give a clear description of the best way he knows of putting his invention into practice, with any alternatives he can think of, aided by drawings. He should not extol the virtues, advantages or value of his invention since these aspects should not appear in a patent specification (see Section 11.2.2).

11.2 Procedure for making a UK Patent application

The procedure for an engineer or scientist to obtain a patent generally follows these lines:

11.2.1 Selection of invention

Among the new features that are devised during a programme of development work, only a relatively small number will usually be worth patenting. The development engineers or scientists will have a fair idea what competitors have been doing; they know which features of their own work were the result of solving difficult problems, and which features are likely to be commercially advantageous. The people best qualified to select the features worth considering for patenting are therefore the development engineers themselves.

11.2.2 First written description

The first description of a selected invention should be written by the inventor himself. A practical example of the apparatus or process should be described. Features that are considered ingenious should be pointed out and compared with the nearest examples, known to the inventor, of apparatus or processes that were in existence before. Sketches or working drawings should be included where possible.

The inventor's write-up need not be at all formal, and need not be anything like a patent specification. Often an ordinary experimental report that has been written in accordance with Chapter 5 will serve the purpose admirably.

11.2.3 Draft provisional specification

The inventor's 'write-up' and drawings should be sent to the company's patent department, or an outside agency; the patent agent then may be able to draft a provisional patent specification from this information alone, or he may find it necessary to discuss the invention with the inventor. However, even if the patent agent does not fully understand it, the inventor's write-up gives him some ground-work which will make a subsequent interview more profitable than if he is merely asked to 'call and discuss some development work we have been doing'.

The inventor is usually asked to approve the draft provisional patent specification as fully disclosing the invention.

11.2.4 Patent application form

The inventor is also required to sign on the back of the patent application form to indicate his assent to making a patent application in the name of the company, in pursuance of the terms of his employment. The inventor's full names and nationality must be stated in the application form, and his name eventually appears at the head of the printed patent specification.

11.2.5 Lodging of the patent application

The patent department lodges the patent application form and provisional specification at HM Patent Office in Chancery Lane, and informs the inventor of the official patent application number and date.

11.2.6 Disclosures after lodging of the patent application

After the United Kingdom Patent Application has been lodged, the company or inventor can, if they wish, disclose the invention (so far as it is described in the provisional specification) freely without endangering the possibility of obtaining valid patents in the United

Kingdom and abroad. However, HM Patent Office does not publish at this stage anything except the inventor's name and the title of the specification. The company or inventor may, if they prefer, keep the invention secret (except for this name and title) during the three years or so while the application is pending at the Patent Office. More often it is found desirable to disclose inventions to customers much sooner than this. In large companies other departments are usually informed of inventions soon after the patent applications have been lodged.

The company may well have a patent interchange agreement in certain technical fields with various manufacturers abroad, in which case copies of the appropriate provisional patent specifications are usually sent to these manufacturers soon after the patent applications have been lodged.

If the invention is of sufficient military importance to be declared an Official Secret, only the appropriate United Kingdom Government Department is informed. The Patent Office itself has a special system for dealing with inventions of this kind.

11.2.7 Decision whether to complete and/or make patent applications abroad

About six months after the patent application date, the department most likely to manufacture the invention is asked for recommendations:

1 Whether the United Kingdom Patent Application should be completed.
2 Whether any corresponding applications should be made in countries abroad.

Information on further developments is also requested.

11.2.8 Drafting and lodging of complete specification

The complete specification is now prepared. This repeats, amplifies and brings up to date the disclosure in the provisional specification, and includes the additional clauses or 'claims' which define the invention, stating exactly what it is that other people are to be forbidden to copy [40].

Obviously specifications vary in length according to complexity. The record patent held in Britain contains over 1,500 pages and nearly 500 separate drawings (UK Patent 1,108,800 granted to IBM).

The complete specification should be lodged at the Patent Office within twelve months from the original patent application date. If any corresponding patent applications are to be made abroad, various forms need to be signed by the inventor. Copies of the complete specification are sent off to patent agents in the countries selected, who translate them when necessary and lodge the applications in their national patent offices within twelve months from the United Kingdom application date.

11.2.9 Developments of invention after lodging complete specification

After the complete specification has been lodged, no further developments in the invention may be added to it; they can only be made the subject of further patent applications.

11.2.10 Examination by the Patent Office

About twelve months after the complete specification has been lodged, the Examiner's report (known as an 'Official Letter' or 'Official Action') can be expected. This gives the result of the Examiner's search for earlier publications of the same invention, and it may raise other objections. The patent department may be able unaided to deal with any objections, or may require assistance from the inventor. A limited time is usually fixed for response to the Patent Office. There may be further Examiner's reports and further responses before the Patent Office finally notifies 'Acceptance'.

11.2.11 Publication and final number

The 'Acceptance' notice gives the final Patent Number and names the 'Publication Date', about two months later, when prints will be made available to the public.

11.2.12 Opposition

For the inventor's purposes, publication can usually be regarded as the end of the procedure on a patent application. Actually, however, there ensues a three-month period during which competitors may enter at the Patent Office an Opposition to the grant of the Patent. The chief grounds for opposition are:

1 That the invention was not devised by the person named in the Patent Application Form, but was obtained from the Opponent.

2 That the alleged invention has been published or used previously, or is clearly obvious in view of what has been published or used previously.

Oppositions are not very popular because the Patent Office cannot investigate fully the important question of 'obviousness' and the procedure is therefore inconclusive and hardly worth the trouble and expense.

When the Opposition period has expired, the actual formal grant of the Patent is issued by the Patent Office. The process of obtaining a patent from the engineer's original report may take as long as two to three years to complete.

11.3 Copyright

The copyright of an author or artist ordinarily arises automatically when the work is created, and no act of registration is necessary for UK protection. But in America and most Dominions copyrights can be registered and failure to do so may involve inconvenience. Copyright usually lasts for at least 50 years. The right to copy a work may be sold outright by the author or artist for a lump sum payment or he may grant one or more licences in exchange for a royalty of so much a copy or of a certain percentage on the receipt from sale of copies.

A typical copyright notice can be seen at the front of this book where the now international sign © is followed by the author's name and date of first publication. This was agreed at the Berne Copyright Convention and the Universal Copyright Convention.

For those engineers and scientists who write books or manuals it is worth noting that under the existing copyright laws a publisher is obliged to pay the author a reasonable royalty and he must publish within a reasonable time and not delay unduly. Finally the publisher is not entitled to publish the book under anyone else's name without the author's consent. In general, then, copyright automatically belongs to the person who creates the book and he is entitled to make use of all forms of exploitation.

Problems concerning copyright occur when judges have to decide whether a similar work is an imitation. A writer or painter cannot claim a breach of copyright merely because someone else has used his idea. On the other hand infringement of copyright cannot be avoided by transferring a book to some other medium. Thus a novel that is produced as a film or a play infringes copyright. Similarly transferring design from two-dimensional presentation to a three-dimensional object is just as much an infringement of the original design as an exact copy. The problem of copyright has been accentuated today because of the wide range of replication equipment available so that cheap copies can be made for dissemination and exploitation.

11.4 Further reading

A Newby, F., *How to Find out about Patents*, Pergamon Press (1967).

B Institution of Mechanical Engineers, leaflets. Part I: *Rights to Inventions;* Part II: *The Ownership of Inventions by Employers;* Part III: *Advice to Inventor-owners of Patents* (1970).

See also References 39 and 40.

12 Copywriting and sales literature

More and more companies are seeking to get better business performance through their sales literature, advertisements, catalogues and other documents. They are finding that it is desirable to have consistency with a distinctive style in their presentations so as to build up a company image. They are also finding that it pays to have these professionally written, expertly designed and quality printed. Such writing is not performed in isolation like a novel or poem. As with a film where, in writing the script, the writer works in conjunction with the producer, director and performers to produce a group effort having an individuality of its own, so with what is commonly called copy–writing.

Corporate identity is now a requirement for many business organisations and has become an important weapon in the battle of the market place. A firm's image can be improved when company policy includes the design of suitable literature, logograms, lettering, etc. [F].

12.1 Publications

While technical publications have to produce a clear factual description of equipment sales literature must also seek to influence people. The writer must try to associate the sales aspects and functional features of the product with certain motivational factors to create an awareness of a need. Too often this element of persuasion is missing from technical sales literature written by engineers who are trained to keep emotion out of their laboratory reports. Every copywriter must keep in mind that buying decisions are made by people and people have feelings.

The end object of copywriting is to encourage people to buy and to do this the writer has to be able to say something interesting about the product or service being offered. While engineers and technologists rarely become copywriters they may materially assist their company's public relations and sales promotion work by providing ideas and details about their creations.

There is of course a wide range of literature through which potential buyers may be reached. Newspapers, technical and trade magazines, advertising, posters, letter and direct mail leaflets and brochures: the scope is extremely wide and technical writers should understand the different varieties of publications. Writing about microwave ovens for the technical press, factual data regarding power, frequency, weight and cooling systems must be given but for the trade press, emphasis must be placed on installation aspects, maintenance, advantages to users over conventional appliances, and so on, whereas for a consumer magazine a newsy article with appropriate illustrations and diagrams should be written laying particular emphasis on cleanliness, speed of use, safety and reliability. In other words, each type of publication needs its own distinctive style of technical writing. But whichever type of publication is being used, it is absolutely essential for the copywriter to relate one or more motivational factors with every technical fact so that people are persuaded to buy.

It is the producer's responsibility to design and manufacture the goods that the public requires and ensure that these are distributed for easy sale. It is the job of company advertising to inform the public of the existence of such goods. Obviously to achieve good advertising in all its forms the copywriter must be informed about the company's designs, manufacturing processes, and distribution channels. Here the technologist can be useful for he should be able to provide factual background information and details of the company's products. It is the copywriter's job to impart information about the company's products in clear, simple, and direct terms using appropriate motivation factors.

12.2 Motivation factors

In sales literature it is the motivation factor that will be the most decisive for it will play on people's needs and desires [41]. At any given time people are motivated by a variety of internal and external factors. The strength or maybe the conflict of each motive will influence the way they see situations, and the actions in which they will engage. People will buy goods to satisfy a need. They may have needs at the purely physical level—such as food, warmth, shelter or security. Many industrial goods, both luxury and domestic, help to satisfy these psychological needs. But there are other needs that may also influence buying habits, such as power, success, prestige, or the need to avoid discomfort, trouble, anxiety, etc. All these factors must be considered

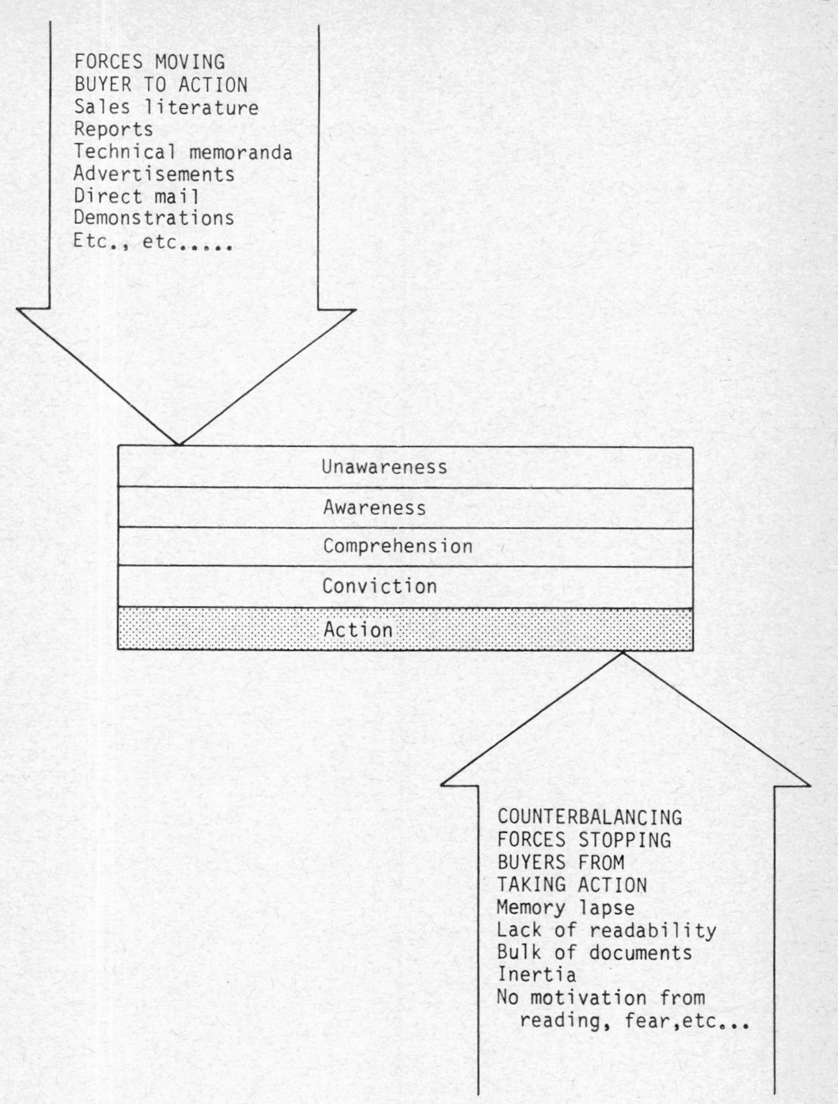

Figure 35 Positive and negative forces concerned in technical writing leading to sales

when compiling technical literature of any sort. Basically, the successful copywriter tries to describe the technical features concisely, clearly and as accurately as possible while at the same time introducing and relating the appropriate motivation factors. Finally he should present the product as a genuine solution to a problem that people have. He tries to move a potential buyer's attitudes from unawareness to awareness and from awareness to comprehension and from comprehension to conviction and so finally to action. This may be looked upon as being like a set of positive forces that are counter-balanced by certain negative forces (see Fig. 35). The negative forces may include fear—fear perhaps that he will lose his money or that the product will not stand up to environmental conditions.

If the product is an aircraft or a 750-MW steam turbine no advertisement or article on its own is going to make an airline or the CEGB buy one so the information presented need not be exhaustive but it has to be telling. Two main points that could be used for such large scale products would be reliability and quietness.

One final point about technical sales literature is worth making. The structure of many sales brochures and manuals can be improved if the fiction writer's artifice of suspense is used. In a novel the structure is arranged to lead up to a climax and the writer adjusts the reader's progress by using suspense. In the same way some sales leaflets use 'expectation'. Here, instead of bold statements about the technical performance of a car fitted with an automatic gearbox in terms of brake horsepower and acceleration, reference is made to the expectation of an exhilarating surge of power when most needed without changing gear.

Good communication between the company's scientists and engineers and its outside agency or public relations department can be a real asset and good technical reports often do provide vital key points for the advertising experts and copywriters.

12.3 Further reading

A Stobo, P., *Copywriting*, Business Publications (1967).

B Hancock, A., *Mass Communication*, Longmans, Green (1968).

C Williams-Thompson, M., *How to Run a PR Campaign*, Pergamon Press (1969).

D Packard, V., *The Hidden Persuaders*, Longmans, Green (1957).

E Williams, R., *Communications*, Penguin (1966).

F Pilditch, J., *Communication by Design: a study in corporate identity*, McGraw-Hill (1970).

See also Reference 39.

13 How to become an effective communicator

In this book an attempt has been made to cover a wide range of technical writing and speaking. That published and spoken technical information often leaves much to be desired cannot be denied. The condition could and will be improved as scientists and engineers seek to become more effective as communicators.

13.1 Communication

A large portion of human behaviour is concerned with transmitting and receiving messages. Indeed in order to survive in our changing environment any organism must develop a capacity for collecting, processing and using information.

While the mathematical theory of communication developed by Wiener and Shannon provides a means for obtaining a precise measure of the information that can be carried in any situation it does not take into account the human factors in communication. In particular the lack of consistency between the encoding and decoding causes considerable difficulty in disseminating technical information.

The fact is that human communication is linked with shared experience but complete mutual understanding is not always possible because of differences in human experiences and abilities to listen as well as speak. The mouth and ear are different organs and it is important to realise that speaking and listening are two separate abilities in the same way that writing and reading are different. In listening a person tries to match the received signal to one he already knows. A listener's first guess is probably derived from various syntactic markers given by a speaker or writer—intonation and inflection of voice, suffixes, looks, gestures, etc., and also in part by the situation. It is important to realise that the listener or reader makes his own private assessment and that learning is assisted when the sender helps the recipient to make his own discoveries.

A human voice cannot generate more than 25 bits* of information

* See Appendix 2

per second yet the capacity of most communication systems far exceeds this figure and the sophistication of many of our channels of speech distribution do not in fact aid comprehension: too often the human factor is neglected.

If this defect is accepted it is the first duty of any communicator to exhibit empathy towards his hearers. He must put himself in their position and see his communications from their point of view. Secondly he must know what he is trying to achieve. With his objectives clearly fixed and a proper appreciation of his audience he will be well on the way to becoming a good communicator.

13.2 Practising the art of rhetoric

There are however one or two further useful points that should be borne in mind by every professional person who wishes to practise the art of rhetoric.

Briefly these are as follows:

1 *Apply organised thinking*—Technical writing is largely a matter of planning and organising. Because this is not seen it is often forgotten—only the good results are appreciated but these stem directly from good organisation.
2 *Acquire a sense of timing*—In both writing and speaking it is important to appreciate when to say something to obtain maximum impact. Sometimes the communicator needs a quick response but on other occasions a gentle lead up to the main point is what is required.
3 *Develop a sense of humour*—Often this ability will carry a communicator through in times of stress and difficulty. Learn to see the ludicrousness of human situations.
4 *Practise again and again*—Whenever you can, practise and never be afraid of your failures. Learn from them, make your own critical assessment of your past efforts and detect where improvements are required. Concentrate on these and seek to correct your faults.
5 *Be honest and sincere*—A communicator should always show complete integrity—there should be a unity, oneness and wholeness about what he has to say and how he says it. Sincere means that there are no cracks in the speaker's or writer's work which are filled with wax to camouflage the weakness.

6 *Use emotional channels*—Remember man communicates by using the emotional, physical and intellectual. The truth is that fact and logically presented argument is only acted on when the receiver is emotionally and physically prepared to act.

If these factors are kept clearly in mind and then practised with the ideas presented in the other chapters of this book there is no reason why the majority of engineers and scientists should not become good technical expositors able to convey information gracefully and securely.

13.3 Further reading

A Carnegie, D., *How to Win Friends and Influence People*, Simon and Schuster, New York (1936).
B Lorayne, H., *Secrets of Mind Power*, Frederick Fell, New York (1961).
C Peale, N. V., *The Power of Positive Thinking*, Prentice-Hall, New York (1952).

14 Remote communications and their integration into the man-machine society

This book is about the communication of technical information. If it had been written 200 years ago the subject matter would be mostly agriculture. It would then make sense to include a chapter on communication with animals for animals were then closely integrated into technology as power sources, means of transport etc.

Today animals have been largely replaced by machines except as sources of protein. Many of these machines carry out far more complex functions than any animal and are far more closely integrated with our way of life. But we still have to tell them what to do, however automated they are. Furthermore we now communicate with other human beings via machines to an extent unthinkable even 100 years ago. Therefore we are interested in the man/machine interface in two different ways: with the machine acting as the communications medium; and with the machine as the recipient of the message.

One need not agree with McLuhan that the medium *is* the message to appreciate that the medium *affects* the message, just as even the best glass affects the light that passes through it. Anyone who doubts that machines have affected the content, as well as the method, of our communications need go no farther than the very concept of interface: this fashionable word, like many others, owes its existence to computers.

14.1 The machine as medium

14.1.1 Printing

It could be argued that communication includes transport in which case the first communications machine was the wheel (or possibly the sail) and the first interface the axle (or the rudder). But if we stick to message communication as such, the first attempt at mechanisation was clearly the printing press; and it probably remains the most important to this day. A whole book could easily be written on the effects it has had on the communication of technical information. Here we cannot devote much space to it since there are many more recent innovations which claim our attention.

However, just in case the reader thinks that after 500 years there cannot be much left to say about printing as a means of communication we shall quickly pinpoint one aspect of Gutenberg's invention which the ever conservative world of Academe does not yet seem to have appreciated. The formal lecture system, which still holds sway throughout education all over the world, was originally nothing more than a method of manual duplicating. When students could not afford textbooks, teachers like Socrates would hold forth in public while a dozen eager beavers like Plato would scribble down what he said. The result was twelve not very good copies but clearly much cheaper than getting a scribe to make a dozen copies in series.

Five hundred years were clearly not enough to make academics realise that this technique became obsolete with the advent of printing. It still remains a slow and inaccurate way of 'transferring the message from the notes of the lecturer into the notes of the student without it passing through the mind of either'.

14.1.2 The telegraph

Semaphore telegraphy dates back at least to the 18th century and with the advent of the electric telegraph the medium achieved an enormous influence on our way of life. It is difficult to realise now that it increased communication speeds by a factor of 30 at a stroke, a far greater increase than the railways achieved in transport. A good case could be made out for the telegraph having opened up the American West, rather than the railways. Certainly without it the U.S. cavalry would not have arrived on time in the last reel!

Telegraphy has had several important lasting effects on communication. First there was the 'telegraph style' which left its mark on journalism, fiction writing and perhaps all human intercourse. If you pay so much a word you tend to choose your words carefully. Unfortunately, the invention of the telephone has undone much of this.

It also became necessary to appoint a specialist man/machine interface in the person of the telegraphist but what was far more important was that people had to start thinking hard about information theory even though the name was not invented for another 80 years. One result was the Morse code; another was the appreciation of the importance of redundancy in language to overcome 'noise'. It is quite an art to word a message so that it will remain intelligible even if one letter in every word is wrong. This is an exercise every writer should attempt.

The telegraph brought world-wide news into every one's home and speeded up the whole pace of civilised life. Although it has now been largely superseded it has left a permanent mark on human society. In the form of the Telex it is still an important means of business communication and it could be more important still if people thought more carefully about the messages they sent.

14.1.3 The telephone

Compared with the telephone, however, the telegraph was small beer indeed. At first sight it is not very clear why this should be so. What can you say over the phone that you cannot say in a telegram? The answer is clear from information theory. The number of words per minute transmitted may not be that much larger but the total amount of information transmitted is nevertheless increased by a factor of several hundred. Not all of this is useful of course but much of it is quite essential. When the chap at the other end coughs you may learn only that he has a cold; but when he hesitates you know there is something fishy in what he says. The human voice carries thousands of overtones, many of which are significant in emotional, if not in technical, terms.

As we have pointed out in Chapter One there is an enormous amount of information we convey, between the lines, as it were; much of it travels in the voice. The telephone has made this available at the speed of light. On the other hand, in the absence of hi-fi telephony, a certain amount of voice distortion does take place and this must be allowed for. People sound colder on the telephone than face to face; part of this is due to the absence of visual gestures which normally go side by side with sound.

Allowance must also be made for various forms of telephone shyness. Some people are better on the phone than face to face but most are slightly afraid of a disembodied voice, especially if it belongs to a stranger. Again they miss the visual gestures which would otherwise put them at ease. It is therefore necessary to cultivate a special telephone manner which takes account of one's own and the other person's diffidence.

For technical information, the telephone is a most unsuitable medium. Technical data contains closely packed facts which are difficult to digest instantaneously and, unless tape-recorded, do not remain on record. Technical terms rely on spelling much more than ordinary words. Also, most technical discussions become clearer when supplemented by sketches.

It is therefore unwise to have important technical discussions on the telephone, however tempting and timesaving this might seem. This situation might change radically when telephone-linked closed-circuit TV becomes readily available. This would also open up an entire new field of remote conference facilities. Just as TV can now link a discussion between speakers in several different towns with no obvious ill effect on their ability to communicate, the TV-phone should make it possible even for a design team to confer and co-operate over long distance.

Again, however, allowances will have to be made for the man/machine interface. Even when complete technical fidelity of transmission is achieved it will be a long time before most people talk as freely to a microphone as to a human face. Otherwise why do even experienced performers need a live audience on TV?

The tape-recorder message taker is a useful modern adjunct to the telephone which can, however, do more harm than good if used as a replacement, instead of an adjunct. Some people in business prefer to work from a permanent record which enables them to reply in their own good time, if ever. This sort of discourtesy is no substitute for personal service and is a good example how modern methods of communication can cause us to be farther apart.

Another diabolical aspect of telephones in business is indifference and discourtesy to strangers. A letter does not know when it is being kicked about from department to department and unless the delay is excessive, the writer does not care. But who has not been infuriated by being shunted about from extension to extension in a large organisation through no fault of his own? Ministries are the worst offenders but large companies have nothing to be proud of. Usually the culprits are juniors who would never be allowed to answer a letter. But they can quite easily put off important callers by discourtesy.

Particularly the switchboard operator who, should always be entirely familiar with who is who and where; and who does what; in the organisation. Why not get a friend to ring up your own switchboard, pretending to be a stranger and asking an awkward question? You might see your whole organisation in an entirely new light.

14.1.4 The mass media

Telegraphy and telephony as well as printing are not, of course, fundamentally changed by being addressed to a mass audience, be it by means of Hertzian waves or newspaper distribution. The content of the message, however, will, as always, have to be adapted to the audience, as will the technique of communication. When using the mass media it is impossible to be sure who the individual recipient will be, what he knows and what he wants to know. This means that a heavier brush will have to be used and poster colours instead of pastel.

When writing for the daily or general press, no scientific knowledge of any kind must be assumed; not even elementary numeracy. If you have ever been put off by someone shooting you a line about Proust or Plato do not expect your general reader or listener-in to attach any meaning whatsoever to the second law of thermodynamics. It's not his job to understand you, it is your job to make everything crystal clear to a scientific idiot. Pretend you are programming a computer: they are quite complex machines but they have to be told before they know.

Otherwise, what has been said about lecturing applies to the mass media even more than to specialist talks for the initiated. To make sure, always try it first on the wife or the schoolboy son.

14.1.5 Television

Television introduces an entirely different order of communication. The information capacity here is as superior to the radio as the telephone is to telegraphy.

There is and will continue to be an increasing development within existing TV systems. The portable TV is now with us, and will be extensively developed and so allow mobile viewing. Also cable distribution of television systems can be linked via computers with a range of services which will allow users to 'dial in' for the information requirements. Some of the possible uses for the dissemination of information will probably be:

a) Wired news, weather and traffic information services.
b) Shopping services, linked to a telephone system, so that goods can be seen and immediately ordered.
c) Educational programmes.

d) 'Demand' information services from libraries and other memory banks.
e) 'Demand' programmes—special TV shows, films.
f) Medical consultation services.
g) Public meeting, discussions, conferences and voting.

Obviously these developments will have large social repercussions and as communication systems may well alter the way people conduct their work. Already Closed Circuit Television (CCTV) is being used for conveying technical information. This can be used for conferences or between departments in the same or different enterprises such as design information being conveyed to production units.

Perhaps the most revolutionary technical development would be the introduction of interactive television. Here the recipient of a programme would respond in certain predetermined ways enabling him to select from items on display or from an advertisement. Such a facility could radically alter the buying habits of individuals and companies.

14.2 The machine as recipient

Communication with machines is not of course new; as has been observed, when the wheel was invented it became necessary to communicate with it via the axle. However, in the narrower sense communication concerns information transferred between intelligences of sorts via symbols such as words, letters or other codes. It arises, therefore, only when a machine can be said to have a rudimentary form of intelligence.

Since intelligence is nothing if not flexible, we must exclude the older form of automatic devices which indeed need a message to tell them what to do; but the message is built-in once and for all.

The purist will tell us that we are on dangerous ground here because magnetic tape is not much more flexible as a means of programming than mechanical dogs and cams. But modern methods of automation are so obviously superior in scope to their predecessors that the man/machine communication interface is likely to become a far greater problem in the future than it has been so far.

14.2.1 Remote control

By way of transition between the two sections of this chapter it is as well to start with the increasing use of machines as media to communicate with machines at the other end: in other words, the control of automatic plant via telephone or closed circuit links. A great deal of this has long been done within large plants, particularly the highly automated process plants. But perhaps the ultimate in this direction has been reached with the ability to use ordinary telephone company lines for sending coded instructions to computers and other automated plant anywhere in this country; and on special occasions half-way across the world.

This is not the place to go into the technical details of how it is made possible; the industrial implications are enormous. One can foresee a time when not only will managers and technologists all over the country be able to co-operate as if they were in the same office but operators will be able to control plant in remote factories while doing another job in their own works. At present this is only done with more or less trouble-free plant such as standby electricity generators but one can easily visualise running numerically controlled machine tools in this way.

As regards actual communication techniques, no new problem is raised, simply a combination of what has been said about machine media and what will be said about talking to machines.

14.2.2 Computers

Computers are the best example on which to study the man/machine interface problem. For what could be more to the point than automatic information handling and processing devices. Basically they are nothing but great adding machines equipped with elementary logic. They have been described as 'idiots' to reassure those who are afraid of a coup d'etat by half a dozen IBM machines.

But it is not as simple as that. The ability to do elementary mathematics and take logical decisions on the basis of it is undoubtedly an important function of intelligence. Just how important one begins to realise when one looks at all the things computers can do, from composing rather poor music to running a factory as well as the average filing clerk could do.

Fortunately, perhaps, they do have important limitations: like Mr. Spock they have no emotions, no sense of humour, no creativity or even imagination. Above all, they cannot learn from experience. Their verbal and visual facility is poor because these senses in humans depend heavily on emotional overtones, as has been pointed out in

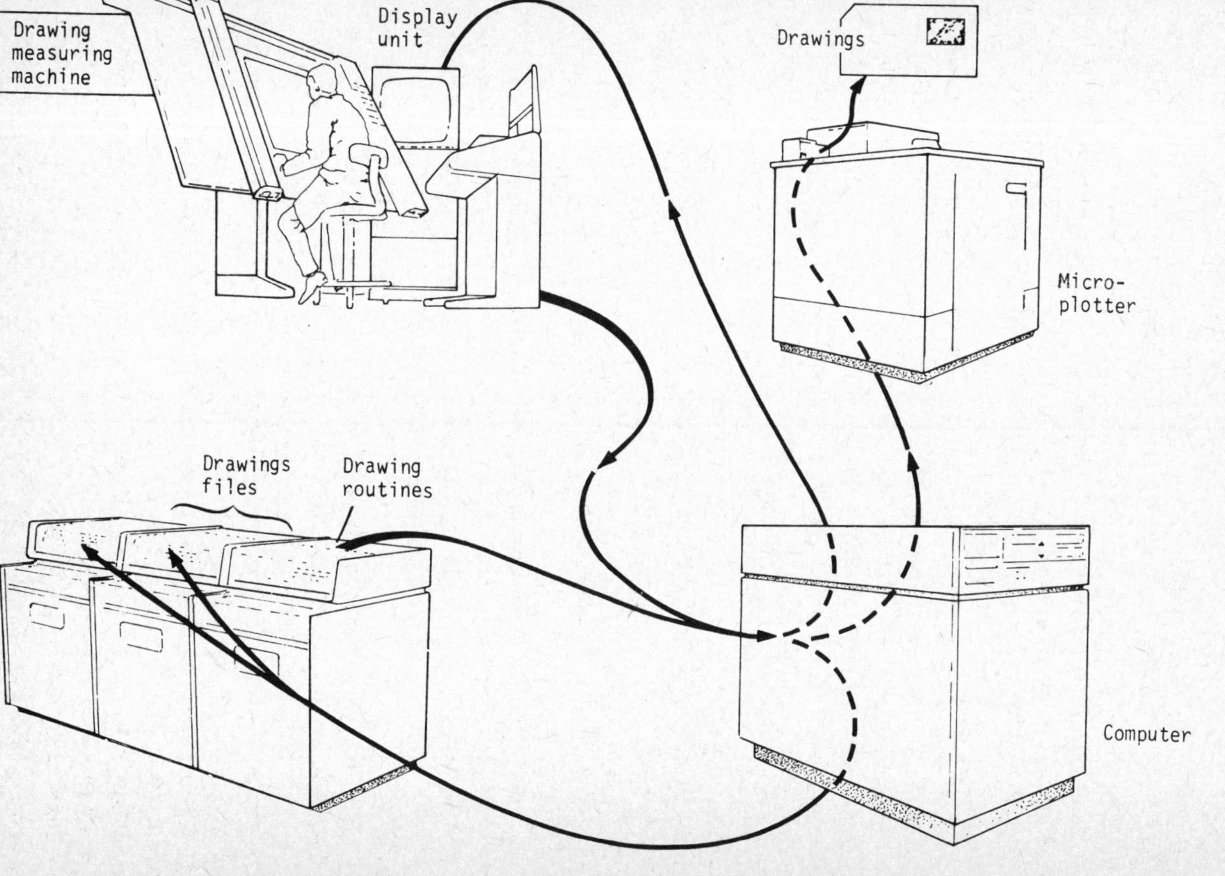

Figure 14:1 Computer produced drawings (on line)

connection with the telephone and TV.

This tells us something about communication with computers and, for that matter with all commercially available automatic machines. They are exceedingly logical and obedient, which we are not. Tell an NC machine tool to cut off its worktable and it will do so. Tell it to check first whether it is cutting the right chunk of metal and it will do that too.

Programmers soon learn to attend to every little detail and anticipate every conceivable eventuality of the kind which a human being, however stupid, would spot for himself. This is valuable training for the programmer. It should enable him to become a first class teacher of humans, production planner and even manager. For to be a good programmer he must supply all the qualities of imagination and foresight that the computer lacks while understanding clearly how the machine 'thinks'.

The subject of computer languages is enormous and complicated, not least because the manufacturers of computers seem to be quite unable to agree on International standards. However, the real problems lie in the interface. The more acceptable a method of communication is to the computer, the less so it is to the human operator or programmer. Ideally the machine wants an input consisting entirely of code numbers, giving it detailed instructions for every switching operation it carries out at the rate of millions per second. But the programmer would like to type out in plain English: 'take this week's tax deductions and update last week's payroll'. In practice various compromises have to be found. Typical computer languages are to be found in Appendices A3 and A4.

Computers and Design: Ideally then when using computers it is necessary to allocate the creativity aspects to the man and any routine and search aspects to the machine. A good example of this is when the computer is used in engineering design. Here a designer may scheme out a design on a drawing board and use a cathode ray tube with a light-pen to add to or modify his design as he proceeds through the design process to develop details of a component. Such a system is diagrammatically illustrated in Figure No. 14.1 and may be used to produce working drawings from a plotter or micro plotter. Thus a total design concept may be stored on the computer. As layout proceeds via graphic displays further retrieval of stored data on Standards and Specifications is called up. Finally when the design process is complete, an entire data bank exists on the product and detail drawings can be produced automatically.

The heart of graphic input/output is a cathode ray tube equipped with a light-pen, which may be used to draw a diagram or to define a spot on the screen by pointing at it. This allows a designer to carry out a dialogue with a digital computer before he instructs it to produce drawings. Here the method of operation involves electronic circuits which are similar to radar tracking devices. Control buttons can be used to indicate what curve is required, e.g. straight line, circle etc. If a hand-drawn line on the VDU is required to be made straight and vertical, then with the light-pen pointing to it and the appropriate button pressed, the stored data would be reprocessed to make the line vertical through the light-pen's centre. Scale may be obtained by specifying the length of one line and all others will be automatically redrawn to suit.

However, a light-pen is not as easy to use as a pencil, for one thing it is a blunt instrument and the working surface is the protective face place of the VDU. Designers using this tool must abandon normal drawing board approaches and adjust to push-button action. Furthermore, the 'window' size is comparatively small and for mechanical engineering work this means that the full drawing has to be built up picture by picture. Expertise at the graphic display demands concentration and fatigue can quickly occur. Light-pen-equipped computer systems have already been used successfully for sketching out shapes, e.g. styling motor cars, where precise models and drawings might take months. Here computer graphics promise to cut down styling exercises to weeks or even days. The same applies to lofting work for aircraft and ships.

In electronic circuit design a computer can provide 'feel' for the designer who can inject signals into the proposed circuit and obtain output response with varying values of components, details of which have been stored. This is a very powerful design tool, giving quick visual display of performance interactions by functional units. Similarly the computer is a very valuable tool for designing printed circuit boards (PCB's) saving many valuable man hours. The same applies to all products built from standard modules. The computer stores part and sub-assembly information which can be speedily retrieved and turned into a design to meet a particular customer's requirements. With complicated shapes, such as turbine blades, it is possible for the designer to set up a profile on the VDU and get performance curves displayed to depict the required characteristics. Such curves may then be modified for aerodynamic reasons.

What we learn from communication with computers is that every message can be taken down to its basic 'bits' of information. Human beings are capable of absorbing vast amounts of knowledge in a lump but there is a limit, depending on the recipient. All communication is a kind of programming in that we wish the recipient to be able to take certain action on the basis of it. If we can speak to computers, we ought to be able to make things clear to far more intelligent human beings, always provided we take the same precaution of feeding it in a language they understand.

14.2.3 Learning machines

If computers are not particularly intelligent because they lack creativjty and the ability to learn from experience, there have been attempts at machines with the latter faculty. It could be argued that creativity is merely serendipidy, the ability to jumble ever new combinations of memory elements and then make a correct value judgment on anything useful that comes up. Judgment, of course, arises from the ability to learn. So that, if we could devise a true learning machine, equipped with a computer, we humans would really have to look to our intellectual laurels.

Learning machines must not be confused with teaching machines, such as programmed tapes (or books, for that matter). These 'program' you with an element of knowledge, then you feed it back in some way and the machine will only go on to the next bit if the feedback was correct. It may even go back to the previous item, if you get the current one wrong more than once. Of course, just as every good teacher must be an able learner of his pupil's reactions, so a teaching machine is in a way a learning machine since it reacts to what it has discovered about the pupil's knowledge. But this is a long way from learning to ride a bicycle.

Teaching machines and programmed books have a lot to teach us all about communication technique. All communication is teaching (or hopes to be) and the method of cutting up knowledge into easily digestible lumps and feeding back the reaction is ideal for all subjects that can be taught without emotion or value judgment. The method has of course all the virtues and vices of the tape recorder: it cannot put you off by sarcasm but it does not provide a great deal of encouragement either. Though a similar device for retarded children has been equipped to pay out Smarties for every correct answer, with great success!

But the real interest of the future lies in true learning machines: something, suitably equipped with sensors, that will learn to ride a bicycle or optimise a process plant in the same way a human being does, simply by doing, watching and thinking, by cancelling false moves and reinforcing correct ones. All you would have to tell it is the result you want. No doubt, it would then tell you that you ought to want a different result!

Primitive machines of this kind have been built; and even programmed into computers. At present the fashion is to develop numerical control that will sense when a machine tool encounters, say, a piece of harder metal, and optimise the cutting speed accordingly. Since we do not yet know exactly how the brain works we can hardly hope to reproduce its major function, learning, in hardware. But the study of both biological and machine intelligence is getting increasingly interlinked and the results of cybernetics are already throwing increasing light on the subject of communication.

For example, computers can now read handwritten letters (up to a point) and recognise spoken words (again very much up to a point). In developing these devices much has been learned about reading and writing that will come in useful in education and communication.

A1 References

1 Marshall McLuhan, *Understanding Media—the Extensions of Man*, Sphere Books (1968).
2 Marshall McLuhan, *The Medium is the Message*, Penguin (1967).
3 Sir Ernest Gowers: *Plain Words*, HMSO (1948); *ABC of Plain Words*, HMSO (1951).
4 Belson, W. A., *An Enquiry into the Comprehensibility of 'Topic for Tonight'*, BBC Audience Research Department Report (1958).
5 Koestler, A., *The Art of Creation*, Hutchinson (1964).
6 Rothman, R. and Perucci, R., 'Career histories of 4000 engineers', *Administrative Science Quarterly*, 15(3), p. 282 (1971).
7 Professor Sir Owen Saunders, 'Communication and teaching', *Chartered Mechanical Engineer*, pp. 19–20 (January 1971).
8 Thompson, E. P. (Editor), *Warwick University Limited*, Penguin Education Books Special (1970).
9 Snow, C. P., *The Two Cultures and the Scientific Revolution*, Reith Lectures, Cambridge University Press (1959).
10 Hudson, L., *Contrary Imagination*, Methuen (1966).
11 Kirkman, A. J., 'The communication of technical thought', *Chartered Mechanical Engineer* (December 1963).
12 Jones, J. C., *Design Methods*, Wiley/Interscience (1970).
13 BSI, *Guide to the Preparation of Specifications*, PD 6112 (1967).
14 BSI, *The Operation of a Company Standards Department*, PD 3542 (December 1959).
15 Binney, R., *British Standards*, Newman Neame Take Home Books (1966).
16 *Conditions of Contracts for Works of Civil Engineering Construction*, Federation Internationale du Batiment et des Travaux Publics.
17 *Value-added Tax*, HMSO (March 1971).
18 Fishlock, D., *Man Modified*, Paladin (1971).
19 Thomas, C. A., *Programmed Learning in Perspective*, Lampon Technical Products Limited (1963).

20 Semler, E. G., 'What's wrong with the technical press', *Chartered Mechanical Engineer* (December 1970).
21 Kapp, R. O., *The Presentation of Technical Information*, Constable (1948).
22 Flesch, R., *The Art of Readable Writing*, Collier-Macmillan, Canada (1962).
23 Semler, E. G., 'It takes two to communicate', *Chartered Mechanical Engineer*, pp. 250–255 (July 1971).
24 *Writers' and Artists' Year Book 1972*, Adam and Charles Black (1972).
25 What Managers Read, OPN3, BIM Publication (1964).
26 Kirkman, A. J., *Business Communications*, paper presented at PERA Symposium on Writing and Speaking Effectively (November 1966).
27 Iveus, M., *The Practice of Industrial Communications*, Business Publications (1963).
28 *Information Systems for Designers*, University of Southampton International Symposium (July 1971).
29 *Technical Information for Engineers—What They Need and What They Get*, CEI and ASLIB Conference (October 1968).
30 Mambert, W. A., *Presenting Technical Ideas: a guide to audience communication*, John Wiley (1968).
31 Heap, H. R., 'Writing and speaking effectively', *The Professional Engineer* (August 1968).
32 *Tips on Talking*, BACIE Booklet (1961).
33 Anstey, E., *Committees—How They Work and How to Work Them*, George Allen & Unwin (1962).
34 Puckey, W., *So You're Going to a Meeting* (1957).
35 Wright, A., *Design for Good Visual Aids*, Studio Vista, London (1970).
36 Taylor, E. A., *Visual Presentation in Education and Training*, Pergamon Press (1966).
37 Young, H. W., 'Maintenance philosophies for the 1970's', Parts I and II, *Industrial and Scientific Communication* (February/March 1970).
38 Whitehouse, F. E., *Publications Estimating by Computer*, BAC Stevenage (1971).
39 *About Patents—Patents a Source of Technical Information*, The Patent Office (1971).

40 Bowler, J. E., 'Patents for designers', Parts I and II, *Journal of Automotive Engineering* (March/April 1971).

41 Banks, J. G., *Persuasive Technical Writing*, Pergamon Press (1966).

A2 The 'bit'—an element of information

It has been pointed out that the extension of any one sense alters the way people think and act and the way they perceive the world around them (1, 2). The dominant organ of sensory and social orientation in pre-alphabet societies was the ear. The alphabet forced a new medium on man causing him to use his eyes for communication. Printing created a popular book and literacy conferred the power of self-improvement. When talking about information today we envisage a wide field of coverage written words, pictures, sounds, signals, perception etc. To measure and compare information carried by various media a common element is required—a 'bit' or binary digit. This is the smallest possible element of information. Engineers in design and production require an enormous number of bits of information to carry out their functions. Some idea of the quantity can be gauged from the fact that an alphabet can be made from 405 bits and that a normal typist works at about 240 characters or approximately 96,000 bits/min. This means a typist works at 1,600 bits. Similarly a designer draughtsman can produce many bits a second when making a drawing for production. On many engineering projects today it is not uncommon to have as many as 50,000 to 100,000 separate drawings, and each drawing may contain many thousands of bits of information. The recipients normally have but two ways of receiving information, by ear and by eye. The ear works at about 100,000 bits and the eye at about 25,000,000 bits though the brain cannot take in information at anything like this speed. Hence the great different in the information capacity of radio and television. But although the media of today allow such quick transfer, the ultimate reception depends upon the human brain which can only comprehend words at about ten per second, i.e. something like 100 bits. A great deal of design/production information is communicated to the eye and control of its comprehension is very difficult to achieve.

Some idea of the information quantities put out today on a world scale can be gleaned from the following:

Sounds and words		
Books	4,000	
Newspapers	1,400,000	
Music scores	1,000	
Gramophone records	22,000	
Radio stations	2,000	
Total	1,429,000	million bits/sec
Sound and vision		
Cinemas	2,000,000	
TV stations	100,000	
Total	2,100,000	million bits/sec
GRAND TOTAL	3,529,000	million bits/sec

A3 Some common computer languages

1 FORTRAN

FORTRAN or FORmular TRANslation is a language composed mainly of statements in a mathematical form, extended and interpreted so as to readily specify and evaluate processes for mathematical equations. It originally contained rudimentary facilities for logical manipulation, input and output, which have since been extended in a fairly powerful way. It is available for all IBM computers in one of several versions, FORTRAN II and IV being most commonly used. A form of FORTRAN is also now implemented on most other manufacturers' computers.

2 ALGOL

A mathematical language devised by an international committee with the intention that it should become an international standard. It has useful facilities for logical manipulation of numbers and characters as well as for arithmetical processes, and provides programs in a fairly readable form. It is widely used in journals to describe programs and methods, for which purpose it is reasonably adequate. It has been implemented on many computers, but IBM and ICL have not adopted it as standard software, preferring to continue maintenance and development of FORTRAN. Input and output facilities have not been standardised and differ for each implementation.

3 COBOL

A COmmon Business Orientated Language, devised by a committee in the United States and adopted as a standard by the United States Government for its data processing installations. Changes are controlled by the CODASYL committee and it exists in several versions, each of which retains previous versions as a subset. It is basically a file-processing scheme, and its facilities for data description, report writing, reclassification of data, arithmetical and logical manipulation and even sorting are either clumsy or non-existent. Nevertheless, it is common to all ICL and IBM machines except the 1130 series and is implemented on many others, and can be used for most aspects of commercial work.

4 NPL (now called PL/1)

A joint IBM-SHARE attempt to combine commercial and mathematical facilities in a single language. This is now available on the IBM 360 series computers and since 1970 has become a general language.

A4 Part of a typical COBOL program

```
COBOL PROGRAM LISTING FOR PART OF RUN 1

PROCEDURE DIVISION
     OPEN INPUT  INPUT-CARDS, STOCK-RECORDS
         OUTPUT  STOCK-HISTORY, SUPPLY-CARDS, LISTINGS.
READ-CARD.  READ INPUT-CARDS, AT END GO TO PROCESS-SUPPLY-DATA.
     IF CODE OR ITEM-NO OR QUANTITY IS NOT NUMERIC,
     STOP 'NON NUMERIC CARD'  GO TO READ CARD.
     MOVE ITEM-NO TO KEY-STOCK.
     READ STOCK-RECORDS, AT END STOP 'ITEM NOT ON FILE',
     GO TO READ-CARD.
     IF GOODS-RECEIVED GO TO C.
     IF DISPATCH GO TO D.
     IF SUPPLY AND QUANTITY TO NO-ON-ORDER, MOVE ITEM-NO TO ITEM.
     MOVE NO-ON-ORDER TO DATA.
WRITE-HISTORY.  WRITE HISTORY-RECORD.
  REWRITE STOCK-RECORD.
     GO TO READ-CARD.
C.   ADD QUANTITY TO NO-IN-STOCK,
     SUBTRACT QUANTITY FROM NO-ON-ORDER
     IF NO-ON-ORDER IS NEGATIVE, MOVE ITEM-NO TO E-LIST-1,
     MOVE NO-ON-ORDER TO E-LIST-2, MOVE SPACE TO CONTRL,
     WRITE LISTING-RECORD FROM E-LISTING AFTER
     ADVANCING CONTRL LINES
     MOVE ITEM-NO TO ITEM, MOVE NO-ON-ORDER TO DATA
     GO TO WRITE-HISTORY
D.   SUBTRACT QUANTITY FROM NO-IN-STOCK.
     IF NO-IN-STOCK IS NEGATIVE, MOVE ITEM-NO TO E-LIST-1.
     MOVE NO-IN-STOCK TO E-LIST-2, MOVE SPACE TO CONTRL,
     WRITE LISTING-RECORD FROM E-LISTING AFTER ADVANCING
     CONTRL LINES, MOVE NO-IN-STOCK TO DATA.
     SUBTRACT QUANTITY FROM NO-ALLOCATED.
     IF NO-ALLOCATED IS NEGATIVE, MOVE ITEM-NO TO E-LIST-1.
     MOVE NO-ALLOCATED TO E-LIST-2, MOVE SPACE TO CONTRL,
     WRITE LISTING-RECORD FROM E-LISTING AFTER ADVANCING
     CONTRL LINES, MOVE NO-ALLOCATED TO DATA.
     MOVE ITEM-NO TO ITEM.
     GO TO WRITE-HISTORY.
 PROCESS-SUPPLY-DATA.
```

A5 Typical print-in/out on a remote terminal using the BASIC program

```
NEW

NE   FILE NAME                          PRO 82
10   LET F = 98.4
20   LET C = (F - 32) X 5/9                         PRINTED IN
30   PRINT  'DEGREES',  'DEGREES'
40   PRINT FAHRENHEIT,  CENTIGRADE
50   PRINT   F       ,     C
60   END RUN
```

```
PRO 82   1047  GEISE  28/03/71

       DEGREES                  DEGREES

       FAHRENHEIT               CENTIGRADE
                                               PRINTED OUT
         98.4                     36.8889

       USED  1.67 SEC
```

A6 Empty words

A6.1 Some phrases and sentences to be avoided when writing reports

Statements as written	*What it really means*
It has long been known that	I haven't bothered to look up the original reference.
. . . accidentally strained during mounting.	. . . dropped on the floor.
. . . handled with extreme care throughout the experiments.	. . . not dropped on the floor.
It is clear that much additional work will be required before a complete understanding.	I don't understand it.
Three of the samples were chosen for detailed study.	The results on the others didn't make sense and were ignored.
These results will be reported at a later date.	I might possibly get round to this some time.
Typical results are shown.	The best results are shown.
Although some detail has been lost in reproduction, it is clear from the original photograph that . . .	It is impossible to tell from the micrograph.
It is generally believed that . . .	A couple of other guys think so too.

A6.2 Words to avoid and words to use

A writer's style may be improved by doing away with verbose prepositions and conjunctions and with empty words and by using Anglo-Saxon words whenever possible.

Here is a list of phrases to be avoided:

Avoid	*Use*
a number of	several
following a similar procedure	like
for the purpose of	for
for the reason that	since, because
from the point of view of	for
in order to	to
in accordance with	to, by, under
in the event that	if
in the region of	near
in the same manner	similarly, like
in terms of	in, for
on the basis of	by
on the grounds that	since, because
with the result that	so that

A great many technical reports contain many of the following empty words:

Avoid	*Use*
accordingly	so
consequently	so
for this reason	so
furthermore	then
hence	so
in addition	besides, also
likewise	and, also
more specifically	for instance, for example
moreover	now, next
nevertheless	but, however
that is to say	in other words
thus	so
to be sure	of course

The verb and preposition combination is often used to cover a new thought or idea, e.g. breakthrough, blackout and rundown. Practically all abstract ideas can be expressed by one of the following words, either by itself or combined with an adverb. If these short Anglo-Saxon idiomatic words are used the writer's meaning generally becomes clearer.

Verbs			*Prepositions*	
bear	go	slip	about	forth
blow	hang	split	across	in
break	hold	stand	ahead	off
bring	keep	stay	along	on
call	lay	stick	apart	out
carry	let	strike	around	over
cast	look	take	aside	through
catch	make	talk	away	together
come	pick	tear	back	under
cut	pull	throw	down	up
do	push	tie		
draw	put	touch		
drive	run	turn		
drop	set	walk		
fall	shake	wear		
get	show	work		
give	skip	—		

A6.3 Instant jargon

Finally avoid jargon in your technical writing, if it has to be used always explain what it really means. This may be difficult since the impression is often given that a jargon generator has been used. If any reader feels at a loss for a suitable piece of jargon he should try this generator which consists of three columns of high sounding words numbered nought to nine:

Column 1	*Column 2*	*Column 3*
0 integrated	0 management	0 option
1 total	1 organisational	1 flexibility
2 systematised	2 monitored	2 capability
3 parallel	3 reciprocal	3 mobility
4 functional	4 digital	4 programming
5 responsive	5 logistical	5 concept
6 optimal	6 transitional	6 time-phase
7 synchronised	7 incremental	7 projection
8 compatible	8 third-generation	8 hardware
9 balanced	9 policy	9 contingency

The procedure is simple. Think of any three-digit number at random. Then select the corresponding buzzword from each column. Put them together and you can sound as if you know what you're talking about.

Take for instance the number 257. Take word two from column one, word five from column two and word seven from column three. You now have 'systematised logistical projection'. You don't know what it means but don't worry, neither do 'they'. Would you prefer 'balanced incremental flexibility'? Possibly 'parallel reciprocal options'? Or maybe 'integrated digital contingency'? How about 'functional third-generation hardware' and 'optimal management mobility'? Now that ought to do the trick.

The buzzword generator provides a thousand different phrase combinations, all of which will give you that proper ring of decisive progressive, knowledgeable authority.

A6.4 Avoid the following common errors

A6.4.1 Excessive use of nouns as adjectives

It is not good style to string together a series of nouns for use as adjectives: the use of more than two in series should be avoided if at all possible. Thus:

> 'The Hinckley Point Power Station fan floor concrete slab',

is better as:

> 'The concrete slab forming the fan floor of the Hinckley Point Power Station'.

and

> 'Diesel engine cylinder-head bolt stresses'

The improved versions may be slightly longer but believe me they are shorter to read.

In these 'wrong' examples the reader is kept in doubt as to which nouns are serving as adjectives and which as nouns. His natural inclination is to expect each succeeding noun to be the subject of the sentence only to be disappointed time and again; this has the effect of hindering the smooth flow of information.

A6.4.2 Compound adjectives

Compound adjectives should be hyphenated. The reader, on being presented with,

'Polythene impregnated resin bonded asbestos sleeve; has to pause to work out which words are paired. To the uninitiated it could mean:

'Polythene-impregnated-resin bonded-asbestos sleeve'.

Although hyphenation is usually unnecessary in simple cases, e.g. high-temperature properties or high-pressure valve, if you conscientiously put them in when it doesn't matter you won't forget when it does!

A common error however is to hyphenate compound phrases that are not compound adjectives, e.g. closely-guarded secret. The hyphen is not needed since 'closely' is an adverb and can only qualify the verb not the noun and therefore no ambiguity can arise. However this would often be written wrongly as 'close guarded secret' in which case the hyphen is necessary.

A7 The mechanism of the voice

When presenting a technical case at a meeting it is most important to realise that how you deliver your main points is as important as the actual subject matter. The voice (and its control) is your primary tool. Some understanding of your voice mechanism is therefore desirable.

The human voice is rather like a reed pipe of an organ. The bellows of the organ is directed through the windpipe to the reed. By means of this air, the reed is kept in vibration. This vibration gives rise to air waves which are interpreted as sound by a listener. The resonator influences the sound. Figure 36 shows the organ pipe and the human body side by side for comparison.

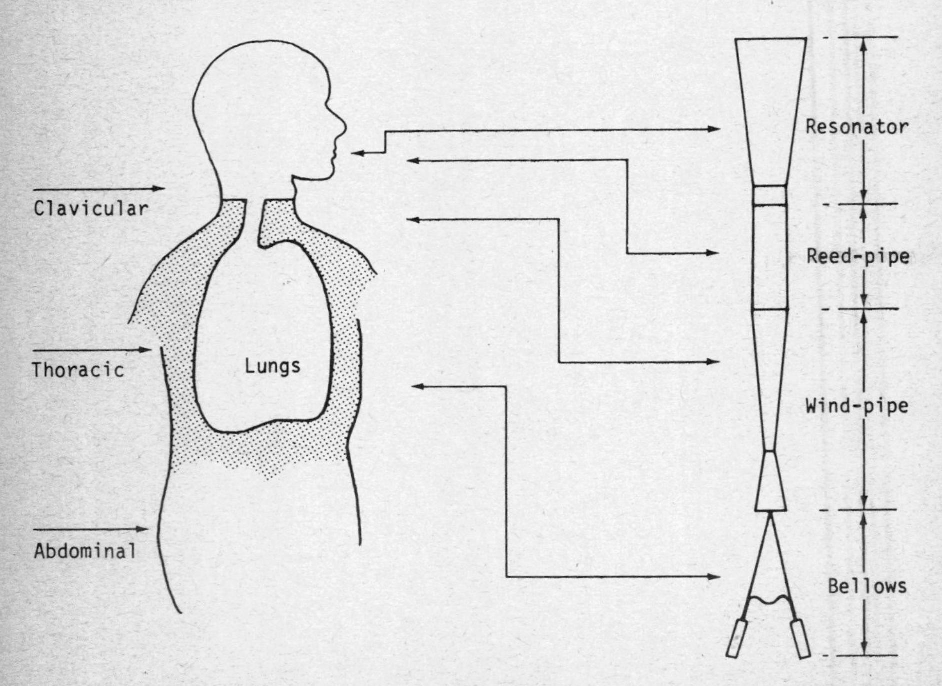

Figure 36 The organ pipe *vis-à-vis* the human vocal instrument

The bellows corresponds to the lungs.
The windpipe corresponds to the trachea.
The reed-pipe corresponds to the larynx with vocal chords.
The resonator corresponds to the organs of articulation, nose, mouth.

Voice tone is moulded by the size and shape of the cavities in the mouth and throat which are regulated by the tongue and the palate.

To speak to an audience requires more breath than normal individual conversation between people. There are three types of breathing:

Clavicular—that which comes from the upper part of the chest.
Thoracic —normal chest breathing.
Abdominal—used in deep breathing.

Abdominal or deep breathing is what is really required for good public speaking. Practice is required to achieve adequate deep breathing and speakers should stand erect with shoulders thrown comfortably back. Air should then be taken into the body and held for a few seconds before commencing to speak.

When presenting your case think rapidly—speak slowly. Speaking slowly, but not too slowly, enables the speaker to duly sound every word so that the hearers can pick up more easily all that is said.

Don't yell or shout but modulate your voice to suit the conditions of the meeting place. It is not necessary to be loud in order to be distinct. Practise using different settings on a tape recorder, so that you can play back your own efforts and learn from errors. Be as natural as you can. When Mark Twain ended one of his soul-stirring lectures he was asked, 'Be that your natural tones of eloquence?', to which he replied, 'Yes'. Avoid the monotonous drawl at all costs, the unvaried continuous sound sends all people to sleep. Again, try not to be too nasal in your speaking but do speak out and never mumble.

A8 Speaking in the style of much scientific writing today

'Daddy, I don't want to go to school today. Why must I?'

'It has been clearly established by several independent investigators that a lack of schooling may lead to a subsequent impairment in an individual's ability to earn money. In addition, other daddies have reported that the particular school for which the present daddy is paying the fees has been found to be a very good one. Another factor which ought to be taken into consideration here is the comparative freedom enjoyed by mummy during the period of daytime when, in view of your absence, there is a necessity to consider only baby and herself.'

'But why must I go to school every day?'.

'The previous statement on this matter has been ignored. It seems likely that you were not listening at the appropriate moment. The present speaker's argument is that in the absence of the educational benefits accruing from attendance at a typical school, failure to learn things may occur, and that this deficiency may, on a later occasion, result in unhappiness, secondary to a limited availability of monetary funds.'

'Daddy, baby's crying. He's always crying.'

'Yes. It has been pointed out that our larval man is particularly vulnerable in this respect. Your observation is in agreement with those reported by both mummy and Uncle Bill. Several other visitors, however, who have studied the phenomenon in other babies, have contested the apparent uniqueness of this aspect of the behavioural pattern of our particular baby.'

'I like Uncle Bill. When is he coming again?'.

'It seems probable, in view of a number of relevant factors, that daddy will achieve visual contact with Uncle Bill during the coming day. This matter will then be considered.

'Daddy, how does your car work?'

'This will now be explained to you in simple terms. . . .'

A9 Report format and cover design

Project Management in Industry

by
F.P. CAMPBELL

BRITISH AIRCRAFT CORPORATION

Guided Weapons Division
BRISTOL AND STEVENAGE

ABC COMPANY
Engineering Report

<table>
<tr><td colspan="3">AIRCRAFT EQUIPMENT DIVISION
Engineering Department</td><td>Salisbury Works</td><td rowspan="2">Report No.

T.1070</td></tr>
<tr><td>Date

18 October 1970</td><td colspan="2">Department Ref.

AED 9621406</td><td>Order No.

5/15379/5</td></tr>
<tr><td>Number
of copies

1
-
-
-
-

1
-
-
-
-</td><td colspan="3">Circulation

<u>Full reports</u>
Mr--

<u>Summary sheets</u>
Mr--

--</td><td>Summary
Reason for report
Description of
equipment concerned
Method of test or
investigation
Results
Conclusions
Recommendations
Enclosures or appendices</td></tr>
<tr><td colspan="4">Title

TESTS FOR RELIABILITY ON THE "ELECTRON" SILICON
CONTROLLED RECTIFIER TYPE AB 100</td><td>Report by

A.N.OTHER</td></tr>
</table>

<u>Summary</u>
The report deals with tests for reliability that have been performed on the "Electron" silicon controlled rectifier type AB 100 ---
----

REPORT No. W/M (IB).n. 5.
AUTHOR. W. N .OTHER

PROJECT PROPOSAL FOR REPLICA METHOD OF STRAIN MEASUREMENT.

SUMMARY

Information on the rapid replica technique of strain measurement was published in 'Engineering' - 6th June, 1958, and a demonstration of the method was seen at the Royal Aircraft Establishment, Farnborough. This consisted in taking two replicas of a good pattern on the surface of a test specimen, first unloaded, and then stressed, and comparing them under a measuring microscope. In this report the method is surveyed and its peculiar advantages over the present means of strain measurement are discussed; the equipment used is described in detail and its cost estimated at £2,000.

NOVEMBER 1958.

2.0 OBJECT OF THE REPORT

The object of this report is to assess the replica method of strain measurement, to survey the equipment required and to make recommendations regarding acquisition of the apparatus.

3.0 INTRODUCTION

3.1 Basic principles of the method

Brief details of the method were published in the June 6th, 1958, issue of 'Engineering' (pp.726-7), although some prior intimation of the method had been gleaned from Company visitors to the Royal Aircraft Establishment at Farnborough, and from a demonstration at the Physical Society's 1958 exhibition in London. The method was developed by, and is in use at the Structures Department of the R.A.E.. A demonstration visit, was, therefore, arranged and the method assessed in practice. This consists in strain measurement by taking a replica of the surface of the material under test which has previously been polished and then marked with a fine rectangular scratch pattern. This replica is produced by impacting onto the prepared surface a platen coated with a low melting point alloy heated to the point where the alloy is in a plastic state. Strains in the two perpendicular directions are derived by comparison of replicas taken in the stressed and unstressed state.

3.2 Availability of the equipment

The equipment as demonstrated is in fully developed prototype form, but the R.A.E. are not proposing to manufacture units for sale. The idea has been submitted to a number of firms, with a view to commercial development. Messrs. Mullard Equipment Ltd., have agreed to investigate the process with a view to putting the apparatus on the market, but it is not expected that a production model will be marketed before 1960.

W/M(1B).n.5

The R.A.E. have made available to us all their working drawings of the gun used for application of the replica platens onto the test area. This is the item on which most development has taken place. The remainder of the equipment is either commercially available or can easily be designed and manufactured.

3.3 Other interest in the company

The equipment had been earlier examined by Mr. Fenton of Applied Mechanics Laboratory, N.E.L., Stafford, and a report (No.S/NE.w.3) was issued. The conclusions regarding the possibilities of the equipment agree generally with those expressed in paragraph 5 of this report.

4.0 EQUIPMENT AND TECHNIQUE

4.1 General

The equipment observed at Farnborough is a completely developed outfit, but in prototype form. Thus, whilst being in working order, the separate units of the apparatus are not arranged as compactly as would be convenient for a laboratory service, and the general design of much of the equipment, such as the gun itself, is based on materials and facilities available in the laboratory rather than in a production shop.

4.2 Preparation of platens

a) Platen Bases: These are 5/8" diameter machined parts and are manufactured from a material similar in coefficient of thermal expansion, to the material of the test area investigated. Only normal machine shop lathes of a suitable size are involved in the manufacture of the platen bases. Figure 1 shows the platen proportions and build-up.

b) Molybdenum Spray: Prior to coating with the replica material, the platens are surfaced by spraying with a thin coating of molybdenum applied by a flame spray gun. Equipment of this nature is available in Whetstone Works and would probably need adapt-

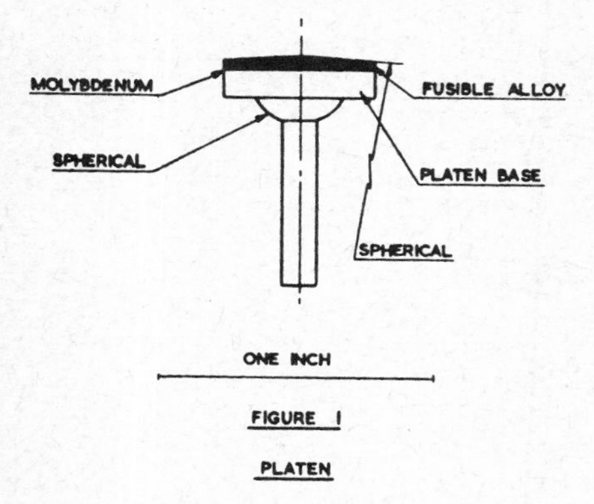

FIGURE I

PLATEN

ation in the form of fixtures for holding the platens in a suitable position. The thickness of the molybdenum coating is merely sufficient to cover the whole area of the platen base.

c) Fusible Metal Spray: The fusible metal is Wood's alloy plus 14% indium. This has an approximate solidus temperature of 50°C and a liquidus temperature of 60°C. The first application of this metal onto the platen base is by a spray fitted with a heater to heat the metal before feeding it into the jets. In order to protect the operator against the toxic metal dust in the atmosphere during the spraying process, the spray nozzle of the gun and a carrier holding a number of platens are inside a closed container fitted with extraction fan and filter. Again in this process, a thin coat of the fusible metal is applied, merely sufficient to cover the molybdenum coating. The heated spray gun was made in R.A.E. laboratories by adapting a gravity fed paint spray of standard manufacture. The heater winding has to be sufficient to raise the temperature of the container, and whole nozzle area, to over 60°C with the spraying air flowing. The gun is handled using an asbestos glove.

d) Fusible Metal Application: The platens are mounted in a heated plate with holes drilled inside it which will contain about 100 platen bases.

The platens are inserted in this plate and when they reach the temperature at which the fusible metal will run, metal is added from a filler rod to form a blob on top of the platen. The platen is then removed from the heating plate and quenched in cold water in order to keep the grain size in the fusible metal to a minimum and so improve the surface condition of the final record.

e) Finishing of Platen: The platen is inserted into a holder which can then be supported on top of a steel ball about 1½" diameter, and swivelled underneath the head of a bench drill which has an end mill mounted in its chuck. By hand guidance, the end of the platen can thus be machined into a spherical surface of about 3" radius. The spherical end of the platen forces out air pockets and thus produces intimate contact of the platen with the test area.

4.3 Preparation of test area

The test area is polished to a mirror finish, i.e. a finish suitable for metallurgical examination. The final finish is produced at Farnborough by jeweller's rouge in a cloth buff suspension. The scratches produced by this buffing should be at 45° to the direction of the lines to be applied to the specimen area for strain measurement.

The reference lines are applied by drawing a pad of rouge paper, across the test area surface, guided by a straight edge and using a light finger pressure. This is done in two perpendicular directions producing a random rectangular scratch pattern on the test area. Heavier reference lines over the test area are then produced by indenting the surface with a sapphire point drawn across the area using the

A10 Communication check-up

A10.1 Before writing or speaking

In order to communicate well, certain questions need answering. It may be useful to apply a kind of check-up using the Rudyard Kipling formula:

'I have six honest faithful friends
They taught me all I knew
Their names are What, Why, When,
How, Where and Who.'

The answers to these names will help you to plan effective and appropriate communication.

What—is the concept idea, piece of information I want to put across? (Is it familiar or entirely new?)

Who—is the reader/listener? (What are the barriers likely to be? Are they the right people?)

Why —do I need to communicate? (To instruct? To inform? To reprimand?)

How —can I interest them from the start?
—can I best communicate? (Speak? write?) Individually or in groups? Direct or through other channels?
—Are they likely to react?

When—is it the best time? (Working hours? Outside the office? Next week? Next month?)

Where—will it go down best? (At the work place? In private?)

A10.2 After writing

Critical examination of written work can be summarised thus:

W = Waffle
I = Inverted sentences
A = Ambiguity
AjN = Adjective noun

J	= Jargon
E	= Elaborate
NP	= New point
R	= Redundancy
SE	= Self-evident statements
IR	= Irrelevancy
C	= Circumlocution
LW	= Logical weakness
OS	= Overlong sentence
TMM	= Too many metaphors (or poor metaphor)

As an example, the reader might like to go through the following extract of a technical article on engine noise. Passages that correspond with the above features are underlined and the appropriate abbreviation placed in the margin.

ENGINE NOISE

A diesel engine embodies a number of mechanisms by
W,R which repetitive pulses of vibration are often occurring and are applied to the cylinder block and crankcase structure. The major sources are the rapid rise
J in pressure in the cylinder upon combustion, piston slap and timing gear rattle (due to crankshaft torsional vibration). Fuel injection pump, injector and valve gear noise can be heard occasionally, but do not affect the overall noise level.

Combustion is often the major source on small
R engines. The pulses of vibration cause the cylinder block and crankcase to vibrate and these in turn transmit the vibration to the thin external crankcase and water-jacket panels. The vibration is also transmitted
AjN to the light section engine covers, such as the sump,
R rocker cover, tapper cover and timing cover. The covers and panels of all the engines investigated have exhibited
J,A mechanical resonances which, when put together, encompass most of the audiofrequency range.

OS,A The flexural motion of these panels can cause noise to be radiated with a sound pressure level proportional to the velocity normal to the panel, when the flexural wavelength of the panel is large in comparison with the wavelength of sound in air at the same frequency. However, the radiation of sound is progressively less efficient at lower frequencies.

AjN, To give an idea of the relative importance of engine
C,J structure noise attenuation and inefficiency of radiation
AjN, from the panels, the one-third octave band level vibrations
A,J of the covers of an engine are plotted as the plane wave
W sound pressure levels equivalent to the vibration velocity of the covers. From the figure the sump (X) is the main
A potential source of low-frequency noise, giving way to the
R tappet covers (Δ) at 1 kHz. At very low frequencies below

A 100 Hz, the engine appears to be rocking, giving high
levels of vibration on the side of the sump and the side of
J,A the rocker cover. However, this mode has a low radiation
efficiency at this frequency and the noise of the air intake
W may be more important. Above 1.6 kHz many of the surfaces
A of the engine are vibrating with similar amplitudes and no
single source can be identified as predominant. The relative
A importance of these covers depends upon the radiation efficiency associated with their modes of vibration. The wavelength of sound in air is 6.75 inches at 2 kHz, which is
smaller than the flexural wavelength of the crankcase panels
(10 inches). The crankcase panels are efficient radiators
C of noise close to the frequency of maximum sensitivity of an
J A-weighted sound level meter (2.5 kHz).

I With some reservations concerning radiation efficiencies
of the various surfaces, the figure shows that all the surfaces of the engine may contribute to the overall noise level,
albeit at various frequencies. The resultant noise of the
engine plotted in the figure was measured 3 feet from the
R,A centre-line of the engine opposite the mid-point on the inlet
side which was mounted on a test bed. The differences between
A the surface vibration spectra and the measured noise includes
some attenuation due to the distance of the microphone from
IR the noise source. Microphones are always difficult to mount
on diesel engines.

When the engine is fitted to a vehicle its noise may
undergo a number of reflections before it reaches an observer
outside the vehicle, and some further attenuation may occur,
R particularly if the engine is well shielded from the observer
as it is in some passenger coach applications.

It is possible to apply noise reduction techniques at any
stage, and broadly the possibilities are:

A 1 Reduce combustion noise by (a) controlling the initial rate
A of injection or by injecting for minimum delay period, (b)
J,A controlling the piston slap by special pistons or by off-

setting the crankshaft or gudgeon pins, and (c) controlling timing gear rattle by smaller clearances between gears or by using a chain or toothed belt drive.

A,LW 2 Improve the structure attenuation by reducing the vibration of all external surfaces of the engine.

LW 3 Place the engine in an acoustic enclosure lined with a sound absorbent to minimise the effects of holes in the enclosure. This will necessitate improved cooling to atone for the added thermal insulation of the enclosure.

The resonances of the tin panel and covers on an engine represent one of the few stages in the path of the noise from source to ear where amplification takes place, and control of these resonances seems to offer a convenient way of reducing noise by perhaps 5 dBA without incurring any obvious penalty in running cost or reliability, but a study of the surface vibration (noise) spectra in the figure shows that several covers or panels could contribute to the noise of the engine in each frequency band, and the summation of these bands with the appropriate weighting determines the overall noise level in terms of dBA. It is pointless, therefore, to set out to treat one or two areas in such a case; the initial aim should be to treat all the covers and then to look for noise sources of secondary importance that may have been masked by the noise from the original covers and panels.

A
OS
W
A
J
R
A

The outer surfaces of the engine may be divided into categories: cast iron panels, cast iron covers, cast aluminium covers, pressed steel covers and pulleys. The treatments tend to follow the construction and material rather than the function of the cover of the engine.

R
A

Cast iron panels may be likened to sound-absorbent sheets acting as drapes in a large lecture hall. Considerable care must be exercised modifying cast iron crankcase and water-jacket-panels, quite small changes may involve considerable alterations and expensive block-machining plant. It is difficult to assess the important modes of vibration of these

PM
LW

	surfaces by theory, and investigations have relied upon exper-
AjN	iment for defining noise-reducing modification. This has take
W,A	the form of a vibration map of both sides of the engine. Thes
J	important modes have occurred on in-line engines:
A,R	1 Bending torsion of the whole crankcase and block, with the
J	'stiff box' of the cylinder block undergoing torsion about
	an axis parallel to the crankshaft, and the crankcase wall
PM,J	bending in sympathy. The first (free-free) mode lies in th
	frequency range 200-630 Hz for the engines tested (8 litre
A	to 2 litres, respectively).
	2 Lateral vibration of the bottom of the skirt on long-
A	skirted crankcases, involving bending of the crankcase
A	panels along a line parallel to the crankshaft. This has
	occurred around 1 kHz on engines tested.
AjN	3 Crankcase panel flexural vibration, where panels opposite
J	adjacent connecting rods move in antiphase. The vertical
W	rotation of the point of attachment of crankcase wall and
	bearing webs may.........

A10.3 Editing

In looking at your own writing it is sometimes useful to have a means of comparing it with the efforts of other people. One way of doing this is to use what has been called the 'Fog' Index. This is easy to calculate by the following procedure:

1. Find the average words per sentence.
2. Count the number of words of three syllables or more per 100 words.
3. Add (1) and (2) and multiply by 0.4 to obtain the 'fog' index.

Try to get an index of about 12 in your writing but remember that a low index does not necessarily mean a clear piece of writing. It will however help to focus your attention on factors of writing style that make reading difficult. Perhaps you would like to consider the following short pieces of edited material and note how they have been improved by the deletion of unnecessary words and reordering.

Mr~~.~~ Smith stated that he had continued with the tests ~~and~~ but ~~made the statement that he~~ had modified the test method ~~with respect~~ ~~to~~ by increasing the temperature~~.~~ ~~The procedure is essentially the same except that the temperature was increased~~ from 80~~°C~~ to 95°C. ~~It was established that 95°C would give~~ He found that this gave the desired effect in ~~a two-week period that previously had taken~~ two weeks instead of two months.

Mr Smith stated that he had continued with the tests but had modified the test method by increasing the temperature from 80 to 95°C. He found that this gave the desired effect in two weeks instead of two months.

~~The manner of placing~~ orders placed on ~~outside~~ sub-contractors ~~indicates that,~~ show that in many cases, the Compan~~ies~~y's ~~established Engineering Staff~~ purchasing procedures are not ~~being~~ followed ~~correctly~~. To avoid confusion and ~~large~~ errors, ~~occurring in the future~~ the attached ~~information is provided to you with the~~ correct procedure is quoted. Further details ~~can be obtained from~~ are given in the Procedures Manual.

Orders placed on sub-contractors in many cases show that the Company's purchasing procedures are not followed. To avoid confusion and errors the attached correct procedure is quoted. Further details are given in the Procedures Manual.

assembles many *regarding*
The report ~~sets out many of the~~ theories and concepts ~~which~~
~~have been developed to deal with~~ ~~M~~echanical ~~B~~alancing prob-
of *so complex* *the*
lems ~~concerned with~~ rotating masses. In a subject, ~~as complex~~
implications of earlier valuble work
~~as this, frequently there are a number of subsidiary researches~~
~~which are recognised for the value when originally presented,~~
~~but during the course of subsequent work their implications~~
of work that follows. This report
are lost in the ~~welter of~~ detail. ~~It is the authors inten-~~
s
~~tion to~~ attempt ~~in this report~~ to bring together ~~the~~ ideas
from *work to give*
~~that have been published in the~~ past ~~and construct therefrom~~
comprehensive *the subject.*
a ~~completed~~ picture of ~~all mechanical balancing related to~~
By its *offers*
~~rotating masses. Because of the~~ nature ~~of~~ the report ~~there is~~
that is *although some*
little ~~particularly~~ new ~~presented in the subject. In many~~
have
~~cases the author has made~~ minor changes and additions ~~to the~~
been made Most theories are presented as restatements or quotat-
~~theories but by and large they are simply rehashes or quotat-~~
ions from the ~~current~~ literature.

this *assembling*
The major contribution of ~~the~~ report is in ~~present-~~
~~ing all assembled information in one place and for suggesting~~
and *all information*
~~methods of~~ ~~inter~~relating ~~it all~~.

The report assembles many theories and concepts regarding mechanical balancing problems of rotating masses. In so complex a subject the implications of earlier valuable work are lost in the detail of work that follows. This report attempts to bring together ideas from past work to give a comprehensive picture of the subject. By its nature the report offers little that is new although some minor changes and additions have been made. Most theories are presented as restatements or quotations from the literature. The major contribution of this report is in assembling and relating all information.

is available for

A special test gear rig ~~has been built for general~~ use in

All

~~the~~ laboratories. ~~It is important to ensure that all people~~

must *their usage time.*

who use the equipment sign ~~for it~~ and enter ~~the appropriate~~

must

~~usage. Should any~~ defects ~~occur in the equipment this must~~

be reported immediately.

A special test gear rig is available for use in laboratories. All who use the equipment must sign and enter their usage time. Defects must be reported immediately.

A10.4 Putting verbs to use

The following are examples of memos and letter writing where improvement can be made by putting the verbs to work.

Instead of: Application of these principles is the best way for any designer to achieve an efficient mechanism.
Say: By applying these principles designers can produce efficient mechanisms.

Instead of: This does have a direct bearing on the possibilities for future designs.
Say: This directly affects future designs.

Instead of: The contract has a written statement to the effect that it should be signed by you.
Say: The contract requires your signature.

Instead of: It is my personal opinion that the sheet piling called for was of the incorrect type.
Say: I think that the wrong type of sheet piling was used.

Instead of: Forward planning has indicated that additional resources of manpower will be needed to meet the scheduled date.
Say: Planning indicates that extra manpower will be required to meet the schedule.

Instead of: This specification is of vital concern to all our design engineers.
Say: This specification vitally concerns all our design engineers.

Instead of: It is the responsibility of the Planning Department to issue working schedules for the production units.

Say: The Planning Department must issue work schedules for the production units.

All Extracts from BS 5261: Part 2: 1976. Recommendations for proof correction and copy preparation.

Table 1. Classified list of marks

NOTE. The letters M and P in the notes column indicate marks for marking-up copy and for correcting proofs respectively.

Group A General

Number	Instruction	Textual mark	Marginal mark	Notes
A1	Correction is concluded	None	/	P Make after each correction
A2	Leave unchanged	- - - - - - under characters to remain	(✓)	M P
A3	Remove extraneous marks	Encircle marks to be removed	X	P e.g. film or paper edges visible between lines on bromide or diazo proofs
A3.1	Push down risen spacing material	Encircle blemish	⊥	P
A4	Refer to appropriate authority anything of doubtful accuracy	Encircle word(s) affected	(?)	P

Group B Deletion, insertion and substitution

Number	Instruction	Textual mark	Marginal mark	Notes
B1	Insert in text the matter indicated in the margin	⋌	New matter followed by ⋌	M P Identical to B2
B2	Insert additional matter identified by a letter in a diamond	⋌	⋌ Followed by for example ◇A	M P The relevant section of the copy should be supplied with the corresponding letter marked on it in a diamond e.g. ◇A
B3	Delete	/ through character(s) or ├──┤ through words to be deleted	∂	M P
B4	Delete and close up	/ through character or ├──┤ through character e.g. charaĉter charaĉcter	∂	M P
B5	Substitute character or substitute part of one or more word(s)	/ through character or ├──┤ through word(s)	New character or new word(s)	M P
B6	Wrong fount. Replace by character(s) of correct fount	Encircle character(s) to be changed	⊗	P
B6.1	Change damaged character(s)	Encircle character(s) to be changed	X	P This mark is identical to A3
B7	Set in or change to italic	─── under character(s) to be set or changed	⊔⊔	M P Where space does not permit textual marks encircle the affected area instead
B8	Set in or change to capital letters	≡ under character(s) to be set or changed	≡	
B9	Set in or change to small capital letters	═ under character(s) to be set or changed	═	

Number	Instruction	Textual mark	Marginal mark	Notes
B9.1	Set in or change to capital letters for initial letters and small capital letters for the rest of the words	≡ under initial letters and ═ under rest of word(s)	≡	
B10	Set in or change to bold type	~~~~ under character(s) to be set or changed	~	
B11	Set in or change to bold italic type	~~~~ under character(s) to be set or changed		
B12	Change capital letters to lower case letters	Encircle character(s) to be changed	ǂ	P For use when B5 is inappropriate
B18.3	Substitute or insert comma	/ through character or ⋏ where required	,	MP
B18.4	Substitute or insert apostrophe	/ through character or ⋏ where required	'	MP
B18.5	Substitute or insert single quotation marks	/ through character or ⋏ where required	' and/or '	MP
B18.6	Substitute or insert double quotation marks	/ through character or ⋏ where required	" and/or "	MP
B19	Substitute or insert ellipsis	/ through character or ⋏ where required	...	MP

Number	Instruction	Textual mark	Marginal mark	Notes
B20	Substitute or insert leader dots	/ through character or ⋏ where required	(...)	M P Give the measure of the leader when necessary
B21	Substitute or insert hyphen	/ through character or ⋏ where required	|-|	M P
B22	Substitute or insert rule	/ through character or ⋏ where required	|—|	M P Give the size of the rule in the marginal mark e.g. |1 em| |4 mm|
B23	Substitute or insert oblique	/ through character or ⋏ where required	(/)	M P

Group C Positioning and spacing

Number	Instruction	Textual mark	Marginal mark	Notes
C1	Start new paragraph			M P
C2	Run on (no new paragraph)			M P
C3	Transpose characters or words	between characters or words, numbered when necessary		M P
C4	Transpose a number of characters or words	3 2 1 | | |	1 2 3	M P To be used when the sequence cannot be clearly indicated by the use of C3. The vertical strokes are made through the characters or words to be transposed and numbered in the correct sequence
C5	Transpose lines			M P

Number	Instruction	Textual mark	Marginal mark	Notes
C6	Transpose a number of lines	3 2 1		P To be used when the sequence cannot be clearly indicated by C5. Rules extend from the margin into the text with each line to be transplanted numbered in the correct sequence
C7	Centre	enclosing matter to be centred		M P
C8	Indent			P Give the amount of the indent in the marginal mark
C9	Cancel indent			P
C10	Set line justified to specified measure	and/or		P Give the exact dimensions when necessary
C11	Set column justified to specified measure			M P Give the exact dimensions when necessary
C12	Move matter specified distance to the right	enclosing matter to be moved to the right		P Give the exact dimensions when necessary
C13	Move matter specified distance to the left	enclosing matter to be moved to the left		P Give the exact dimensions when necessary
C14	Take over character(s), word(s) or line to next line, column or page			P The textual mark surrounds the matter to be taken over and extends into the margin
C15	Take back character(s), word(s), or line to previous line, column or page			P The textual mark surrounds the matter to be taken back and extends into the margin
C16	Raise matter	over matter to be raised under matter to be raised		P Give the exact dimensions when necessary. (Use C28 for insertion of space between lines or paragraph in text)

Number	Instruction	Textual mark	Marginal mark	Notes
C17	Lower matter	over matter to be lowered under matter to be lowered		P Give the exact dimensions when necessary. (Use C29 for reduction of space between lines or paragraphs in text)
C18	Move matter to position indicated	Enclose matter to be moved and indicate new position		P Give the exact dimensions when necessary
C19	Correct vertical alignment			P
C20	Correct horizontal alignment	Single line above and below misaligned matter e.g. misaligned		P The marginal mark is placed level with the head and foot of the relevant line
C21	Close up. Delete space between characters or words	linking characters		MP
C22	Insert space between characters	between characters affected		MP Give the size of the space to be inserted when necessary
C23	Insert space between words	between words affected		MP Give the size of the space to be inserted when necessary
C24	Reduce space between characters	between characters affected		MP Give the amount by which the space is to be reduced when necessary
C25	Reduce space between words	between words affected		MP Give amount by which the space is to be reduced when necessary
C26	Make space appear equal between characters or words	between characters or words affected		MP

Number	Instruction	Textual mark	Marginal mark	Notes
C27	Close up to normal interline spacing	(each side of column linking lines)		MP The textual marks extend into the margin
C28	Insert space between lines or paragraphs		——(or)——	MP The marginal mark extends between the lines of text. Give the size of the space to be inserted when necessary
C29	Reduce space between lines or paragraphs		——) or (——	MP The marginal mark extends between the lines of text. Give the amount by which the space is to be reduced when necessary

A12 Invention and the mis-use of words

Whilst the English language is admittedly a live one and new words have to be invented from time to time to convey an exact meaning, there is rarely any necessity to do so. If you have to coin a word be sure to define it clearly so that all readers understand what is meant when it is used. We can't all claim Lewis Carroll's immunity:

> 'When I use a word Humpty-Dumpty said in a rather scornful tone
> It means just what I want it to mean, neither more nor less'.

Recently, however, two words have been invented to cope with our rapidly changing technology. Both were coined by H. P. Jost—*tribology* and *terotechnology*. 'Tribology' comes from the Greek *tribo* 'to rub' and hence tribology means the study of rubbing surfaces and is therefore used to cover the total field of lubrication and wear. 'Terotechnology' means 'the design of plant, machinery and equipment for minimum maintenance, replacement and removal and the feedback of installation, commissioning and operating experience.' This word is also derived from the Greek—*tero* meaning 'to take care of', so, literally terotechnology stands for the study of taking care of objects, e.g. machines.

Both these terms are useful because they seek to convey a systems approach to multidisciplinary engineering and management. In the same way the term 'overkill'—meaning that our engineered weapon systems now provide a destructive capacity considerably beyond what is required to destroy all mankind—has been coined to cover the effect of the total military technological system.

When the American mathematician Norbert Wiener wrote his book *Cybernetics, or control and communication in the animal and the machine* he described how he had coined the word to designate a group of studies that had not at that time fully taken shape—deriving it, he said, from the Greek work *kybernetes* which literally means 'steersman'. He was apparently unaware that *kybernetike* had been used by Plato and imagined that the term was a new one. So sometimes it is possible to think we have invented a new word which has really been used before.

Sometimes it is permissible in a report to refer to 'so and so's' factor but authors of technical reports should be careful of putting their own name to a formula or law unless they are absolutely sure that no one else has discovered or used it before. While Reynolds number is accepted and used now it was not always so and in fact only came into currency after Osborne Reynolds had died, likewise Maxwell's Demon.

Following the transatlantic lead there is unfortunately, a tendency on occasions to make a verb of an existing adjective or noun by adding the suffix *-ate* or *-ise*; this new verb by easy transformation then becomes a new noun through further extension of *-ate* or *-ise* into -ation or *-isation*, as the case may be.

To take an extreme example the adjective *rugged*, meaning 'rocky' or 'craggy', has at some time been mis-applied in engineering literature in the sense of 'robust'; from this by easy stages we have had thrust upon us *ruggedise* and *ruggedisation*. Any moment now the present writer expects the cycle to be re-started and to come across *ruggedisationise*, and, for a similar reason, *miniaturisationise*!

Pressurisation is another example. The word *pressure* was, some years ago, extended to *pressurise* in order to convey the meaning of 'make a structure capable of withstanding a difference of pressure on its walls'. *Pressurisation* inevitably followed, but should *not* be used, as it sometimes is, for the act of applying fluid pressure to a tank or the like.

Instrumentation is a word which is often mis-used. In its original sense it meant the arrangement of music for several instruments in an orchestra but nowadays the word is also commonly used to denote the art of using measuring instruments. It should not, however, be used as a heading for the list of instruments used in an experiment or test, nor should *to instrument* or *to instrumentate* (!) be used to denote the provision of instruments.

Malfunction is another doubtful word, although officially blessed by use in many Government publications! As a verb it is, perhaps, just acceptable; 'to malfunction' can be interpreted literally as 'to operate badly'. As a noun, however, it should be avoided, as likewise *malfunctioning.*

Differential is an adjective, such as 'differential gear' or 'differential winding'. It should not be used as a noun in place of *difference*, as in 'pressure differential', although 'differential pressure' is acceptable.

Ambient, meaning 'surrounding', is likewise an adjective that is frequently, but incorrectly, used as a noun, e.g. 'the ambient'. Reference should always be made to the noun that it qualifies (e.g. 'ambient temperature' or 'ambient pressure').

Modulate is a verb meaning 'to vary'. It should not be used on its own, as it sometimes is, in the sense of 'damp down oscillations'. You can modulate amplitude and phase and in electronics engineering it is possible to talk about amplitude modulation as well as frequency and phase modulation.

Dissemble is a verb (actually meaning 'to hide under a false appearance') which is sometimes wrongfully used for the reverse of *assemble;* to convey this latter meaning *dis-assemble*, with a hyphen, may be used, but a better word is *dismantle.*

Call-up (shades of conscription in two World Wars!) should not be used in the sense of to 'call for' or 'require' tests, drawings or material parts.

Attenuation—this means 'thinning out', not 'reducing', and should not be applied to structures (see Appendix 10).

In case of any doubt as to the correct meaning of a word, reference should always be made to a dictionary accepted as standard, e.g. *The Concise Oxford Dictionary* and *Chamber's Technical Dictionary*, whilst Fowler's *Modern English Usage* is of value when there is similar doubt on a point of grammar or syntax. But words, of course, change their meaning. In engineering reports one often sees the word *sophisticated* as in 'a more sophisticated system was designed . . .'; the dictionary meaning of sophisticated is still 'mislead, deprive, adulterate'. For this reason it should be appreciated that dictionaries are a collection of precedents, rather than official code books of meaning. Often old words take on a new meaning in times of war or when scientific inventions are popularised. They they pass into national currency; examples are blitz and sound barrier. For the correct use of technical terms reference should be made to the appropriate British Standard, e.g. for electrical work BS204 or BS205.

A13 Procedure manual extract: definitions of required format and contents for a test specification

LIST OF CONTENTS

This manual describes the format and content that these documents shall follow in order that uniformity is maintained. It is based in principle on Manual 5. Where information for a particular unit or piece of hardware is not called for or required, the relevant paragraph shall be endorsed to that effect.

1 PREPARATION

1.1 Paper

These documents shall be prepared on A4 paper. Both draft and frozen copies shall be reproduced on one side only of each sheet

1.2 Typescript layout

The security classification of each document shall be centred at the top and bottom of each page. The disposition of the text shall be such as to leave a clear margin of not less than one inch on the binding edge and not less than half an inch on the remaining edges. The document shall be divided as appropriate into parts, sections, clauses, subclauses and sub-subclauses. Each part, section and clause shall be numbered in Arabic numerals and suitably headed. Subclauses shall be lettered in bracketed lower case type, thus (a) and may, where desirable, be suitably headed. Sub-subclauses shall be numbered in bracketed lower case Roman numerals thus (iv). There shall be no further division.

2 THE SPECIFICATION

2.1 The Specification is a mandatory design document and as such it is the responsibility of the Production Department to ensure that it is produced, approved and maintained. It will identify the equipment and specify, in a concise fashion, precise characteristics and will be a statement of performance in fundamental terms. It defines the essential parameters of an equipment and is a direct interpretation of the applicable Design Specification.

2.2 Duplication of information contained in other frozen documents must be avoided.

2.3 As the documents will be called up on the relevant Item Lists and frozen under routine procedure and as it becomes mandatory once frozen, the author should avoid reference to specific test equipment requirements and confine the text to establishing overall parameters of equipment to be tested.

3 SPECIFICATION SUBHEADINGS

3.1 Definition
All figures and limits specified in the specification shall be absolute values.

3.2 Introduction
The introduction should also include a general description of the assembly to be tested and any special conditions not clearly defined in drawings, etc.,e.g. in the case of the Missile Overall Test in the Magazine - the missile is complete with motor but must not be fired.

3.3 External services
A section shall be provided that details the various services required to activate and operate the equipment to be tested.

3.3.1 General
Power supplies
Air supplies
Oil supplies

3.3.2 Stimuli
Signals

3.4 Performance
Each unit or piece of hardware shall be individually identified and the parameters specified.

3.5 Special conditions

3.5.1 A section shall be provided so that the Design Department can add further information, such as other documents that are required for reference.

3.5.2 Warning safety hazards
(a) Reference will be made to specific safety hazards that may arise in testing, i.e. radiation, explosive, physical (for moving parts), high voltage, etc.
(b) Where special precautions are necessary these should be specified or reference made to suitably associated specifications.

A14 Metrication and technical writing

Basic SI units

Physical quantity	*Name of unit*	*Symbol for unit*
Length	metre	m
Mass	kilogramme	kg
Time*	second	s
Electric current	ampere	A
Thermodynamic temperature	kelvin	K
Luminous intensity	candela	cd

Symbols for units do not take a plurai form.

* *Note:* Although 's' is the SI abbreviation for 'second' many people still prefer to use 'sec' to avoid any possible ambiguity.

Supplementary units

These units are dimensionless.

Physical quantity	*Name of unit*	*Symbol for unit*
Plane angle	radian	rad
Solid angle	steradian	sr

Derived SI units with special names

Physical quantity	*Name of unit*	*Symbol for unit*	*Definition of unit*
Energy	joule	J	$kg\,m^2\,s^{-2}$
Force	newton	N	$kg\,m\,s^{-2} = J\,m^{-1}$
Power	watt	W	$kg\,m^2\,s^{-3} = J\,s^{-1}$
Electric charge	coulomb	C	$A\,s$

Physical quantity	*Name of unit*	*Symbol for unit*	*Definition of unit*
Electric potential difference	volt	V	$kg\,m^2 s^{-3} A^{-1} = J A^{-1} s^{-1}$
Electric resistance	ohm	Ω	$kg\,m^2 s^{-3} A^{-2} = V A^{-1}$
Electric capacitance	farad	F	$A^2 s^4 kg^{-1} m^{-2} = A s V^{-1}$
Magnetic flux	weber	Wb	$kg\,m^2 s^{-2} A^{-1} = V s$
Inductance	henry	H	$kg\,m^2 s^{-2} A^{-2} = V s A^{-1}$
Magnetic flux density	tesla	T	$kg\,s^{-2} A^{-1} = V s\,m^{-2}$
Luminous flux	lumen	lm	cd sr
Illumination	lux	lx	$cd\,sr\,m^{-2}$
Frequency	hertz	Hz	cycles per second
Customary temperature, *t*	degree Celsius	°C	$t/°C = T/°K - 273.15$

Fractions and multiples

Fraction	*Prefix*	*Symbol*	*Multiple*	*Prefix*	*Symbol*
10^{-1}	deci	d	10^2	deka	da*
10^{-2}	centi	c	10^3	hecto	h*
10^{-3}	milli	m	10^6	kilo	k
10^{-6}	micro	μ	10^9	mega	M
10^{-9}	nano	n	10^{12}	giga	G
10^{-12}	pico	p		tera	T
10^{-15}	femto	f			
10^{-18}	atto	a			

* To be restricted to instances where there is a strongly felt need, such as may be experienced in the early days of metrication, in favour of the centimetre as the unit of length in certain biological measurements.

Compound prefixes should not be used, e.g. 10^{-9} metre is represented by

1 nm *not* 1 mμm

The attaching of a prefix to a unit in effect constitutes a new unit, e.g.

$$1\ km^2 = 1\ (km)^2 = 10^6 m^2$$
$$not\ 1\ km(m^2) = 10^3 m^2$$

Where possible any numerical prefix should appear in the numerator of an expression.

Examples of other derived SI units

Physical quantity	*SI unit*	*Symbol for unit*
Area	square metre	m^2
Volume	cubic metre	m^3
Density	kilogram per cubic metre	$kg\,m^{-3}$
Velocity	metre per second	$m\,s^{-1}$
Angular velocity	radian per second	$rad\,s^{-1}$
Acceleration	metre per second squared	$m\,s^{-2}$
Pressure	newton per square metre	$N\,m^{-2}$
Kinematic viscosity diffusion coefficient	square metre per second	$m^2\,s^{-1}$
Dynamic viscosity	newton second per square metre	$N\,s\,m^{-2}$
Electric field strength	volt per metre	$V\,m^{-1}$
Magnetic field strength	ampere per metre	$A\,m/^{1}$
Luminance	candela per square metre	$cd\,m^{-2}$

Units to be allowed in conjunction with SI

Physical quantity	*Name of unit*	*Symbol for unit*	*Definition of unit*
Length	parsec	pc	$30.87 \times 10^{15}m$
Area	hectare	ha	$10^4\,m^2$
Volume	litre	l	$10^{-3}\,m^3 = dm^3$
Pressure	bar	bar	$10^5\,N\,m^{-2}$
Mass	tonne	t	$10^3\,kg = Mg$
Kinematic viscosity, diffusion coefficient	stokes	St	$10^{-4}\,m^2\,s^{-1}$
Dynamic viscosity	poise	P	$10^{-1}\,kg\,m^{-1}\,s^{-1}$
Magnetic flux density (magnetic induction)	gauss	G	$10^{-4}\,T$
Radioactivity	curie	Ci	$37 \times 10^9\,s^{-1}$
Energy	electronvolt	eV	$1.6021 \times 10^{-19}\,J$

The common units of time, e.g. hour and year, will persist, and also, in appropriate contexts, the angular degree.

Until such time as a new name may be adopted for the kilograms as the basic unit of mass, the gram will often be used, both as an elementary unit (to avoid the absurdity of mkg) and in association with numerical prefixes, e.g. μg.

Examples of units contrary to SI with their equivalents*

Physical quantity	*Unit*	*Equivalent*
Length	Ångström	10^{-10}m
	inch	0.0254 m
	foot	0.3048 m
	yard	0.9144 m
	mile	1.609 34 km
	nautical mile	1.853 18 km
Area	square inch	645.16 mm^2
	square foot	0.092 903 m^2
	square yard	0.836 127 m^2
	square mile	2.589 99 km^2
Volume	cubic inch	1.638 71 $\times 10^{-5} m^3$
	cubic foot	0.028 3168 m^3
	UK gallon	0.004 546 092 m^3
Mass	pound	0.453 592 37 kg
	slug	14.5939 kg
Density	pound/cubic inch	2.767 99 $\times 10^4$ kg m^{-3}
	pound/cubic foot	16.0185 kg m^{-3}
Force	dyne	10^{-5} N
	poundal	0.381 255 N
	pound force	4.448 22 N
	kilogramme-force	9.806 65 N
Pressure	atmosphere	101.325 k N m^{-2}
	torr	133.322N m^{-2}
	pound(f)/sq. in.	6894.76 N m^{-2}
Energy	erg	10^{-7} J
	calorie (I.T.)	4.1868 J
	calorie (15°C)	4.1855 J
	calorie (thermochemical)	4.184 J
	BTU	1055.06 J
	foot poundal	0.042 140 1 J
	foot pound (f)	1.355 82 J
Power	horse power	745.700 W
Temperature	degree Rankine	6/9K
	degree Fahrenheit	$t/°F = [(9T/5)/°C] + 32$

* Fuller lists are to be found in the National Physical Laboratory's *Changing to the metric system* (Anderton and Brigg), HMSO, London (1966).

Index

NOTES

NOTES

NOTES

NOTES

NOTES

NOTES

NOTES

NOTES

NOTES

NOTES

NOTES

NOTES

NOTES

NOTES

NOTES

NOTES

NOTES

NOTES

NOTES

NOTES